T0331448

Telecommucations in Disaster Areas

RIVER PUBLISHERS SERIES IN COMMUNICATIONS

Volume 12

Consulting Series Editors

MARINA RUGGIERI
University of Roma "Tor Vergata"
Italy

HOMAYOUN NIKOOKAR
Delft University of Technology
The Netherlands

This series focuses on communications science and technology. This includes the theory and use of systems involving all terminals, computers, and information processors; wired and wireless networks; and network layouts, procontentsols, architectures, and implementations.

Furthermore, developments toward new market demands in systems, products, and technologies such as personal communications services, multimedia systems, enterprise networks, and optical communications systems.

- Wireless Communications
- Networks
- Security
- Antennas & Propagation
- Microwaves
- Software Defined Radio

For a list of other books in this series, see final page.

Telecommunications in Disaster Areas

Editor

Nicola Marchetti

Center for TeleInFrastruktur, Aalborg University, Denmark

River Publishers
Aalborg

ISBN 978-87-92329-48-6 (hardback)

Published, sold and distributed by:
River Publishers
P.O. Box 1657
Algade 42
9000 Aalborg
Denmark

Tel.: +45369953197
www.riverpublishers.com

Cover picture by Alexis Kwasinski

All rights reserved © 2010 River Publishers

No part of this work may be reproduced, stored in a retrieval system, or transmitted in any form or by any means, electronic, mechanical, photocopying, microfilming, recording or otherwise, without prior written permission from the Publisher.

Contents

1

Introduction

Nicola Marchetti

Center for TeleInFrastruktur, Aalborg University, Denmark;
e-mail: nm@es.aau.dk

Disasters occur in short time periods and are usually unexpected, leaving in their wake large numbers of casualties and severe infrastructure damages. These disasters can be due to natural causes (earthquakes, fires, floods, hurricanes, epidemics or combinations thereof) or manmade (industrial accidents, terrorism and war).

Essential communications breakdown is one of the common characteristics of all disasters. The partial or complete failure of telecommunications infrastructure leads to preventable loss of life and damage to property, by causing delays and errors in emergency response and disaster relief efforts. Despite the increasing reliability and resiliency of modern telecommunications networks to physical damage, the risk associated with communications failures still remains serious because of growing dependence upon these tools in emergency operations [1].

Coordinated relief to the affected areas needs to be given as soon as possible, so to minimize further nefarious effects. In such scenarios it is vital that communications between interested parties, i.e. relief and security groups, are established as quickly and as easily as possible, ideally in a plug & play or zero configuration fashion. The acknowledgment that infrastructure-based networks in such deployment areas may be destroyed raises the need for new alternatives and communication paradigms, ideally infrastructure-less, and for decentralized wireless technologies.

N. Marchetti (ed.), Telecommunications in Disaster Areas, 1–9.
© 2010 *River Publishers. All rights reserved.*

1.1 Information Management and Communication in Emergencies and Disasters

Information is the most valuable commodity during emergencies or disasters, since it is what is needed in order to make decisions. In particular, information is necessary for rapid and effective assistance to the areas affected by a disaster. Information needs to be assessed and is the basis for coordination and decision making in emergency situations. It is also essential for after-action analysis, evaluation, and lessons learned. Moreover, public and social communication and media relations have become key elements in efficient emergency management. Technical operations in highly charged political and social situations must be accompanied by good public communication and information strategies that take all stakeholders into account [2].

The following aspects of information are important in the context of emergencies and disasters [2]:

- During an emergency, timely and transparent production and dissemination of information generates trust and credibility.
- Information in emergency or disaster situations comes from many sources; it represents different points of views and serves a wide and diverse range of interests and needs.
- The participation and effectiveness of national and international actors will be beneficial to affected populations to the extent that they have precise, timely, and relevant information.
- The challenges are related to how communication and information management can contribute to a more effective and timely response, and therefore to saving lives, and how this can lessen the impact of disasters and emergencies and improve the quality of life of affected populations.

If the above-mentioned communication measures and expertise are to be valued in the context of disaster management, all necessary technical and human resources must be made available, as well as political backing from health and disaster management authorities. Communication measures and the teams of people responsible cannot be improvised during an emergency; they require ongoing preparation and planning [2].

1.2 Telecommunications for Disaster Prevention and Recovery

Highlighting the role of telecommunications for humanitarian assistance, the former United Nations Secretary General, Kofi Annan said [4]:

> Humanitarian work is one of the most important, but also one of the most difficult tasks of the United Nations. Human suffering cannot be measured in figures, and its dimensions often surpass our imagination, even at a time when news about natural and other disasters reaches every corner of the globe in next to real time. An appropriate response depends upon the timely availability of accurate data from the often remote and inaccessible sites of crises. From the mobilization of assistance to the logistics chain, which will carry assistance to the intended beneficiaries, reliable telecommunication links are indispensable.

Telecommunications are critical at all phases of disaster management. Using a variety of sources that include telecommunications satellites, radar and telemetry, and meteorology, remote sensing for early warning is made possible. Before disasters strike, telecommunications can be used as a conduit for disseminating information on the impending danger, thus making it possible for people to take the necessary precautions to mitigate the impact of these disasters. The first important steps towards reducing disaster impact are to correctly analyze the potential risk and identify measures that can prevent, mitigate or prepare for emergencies. ICT can play a significant role in highlighting risk areas, vulnerabilities and potentially affected populations by producing geographically referenced analysis through, for example, a geographic information system (GIS). The importance of timely disaster warning in mitigating negative impacts can never be underestimated. For example, although damage to property cannot be avoided, developed countries have been able to reduce loss of life due to disasters much more effectively than their counterparts in the developing world. A key reason for this is the implementation of effective disaster warning systems and evacuation procedures used by the developed countries, measures which are not adopted in the developing world [3].

A *warning* can be defined as the communication of information about a hazard or threat to a population at risk, in order for them to take appropriate actions to mitigate any potentially negative impacts on themselves, those in their care, and their property. The occurrence of a hazard might not necessarily result in a disaster, indeed while hazards cannot be avoided, their negative impacts can be mitigated. The goal of early public warning is to ensure to the greatest extent possible that the hazard does not eventually become a disaster. Such warnings must be unambiguous, communicate the risks succinctly and provide the necessary guidance. The success of a warning can be measured by the actions that it causes people to take, such as evacuation or avoiding at-risk areas. To facilitate an effective warning system, there is a major need for better coordination among the early warning providers as well as those handling logistics and raising awareness about disaster preparedness and management. While disaster warnings are meant to be a public good, they are often most effectively delivered through privately-owned communication networks and devices. There are many new communication technologies that allow warning providers not only to reach the people at risk but also to personalize their warning message to a particular situation. It is important to note that *disaster warning is a system*, not a single technology, constituting the identification, detection and risk assessment of the hazard, the accurate identification of the vulnerability of a population at risk, and finally, the communication of information about the threat to the vulnerable population in sufficient time and clarity so that they can take action to avert negative consequences. The latter underscores the importance of education and creating awareness in the population so that they may respond with the appropriate actions [3].

When a disaster eventually strikes, by the above arguments, the coordination of relief work by national entities, as well as the international community is made possible. Recently, this was evident in Grenada where Hurricane Ivan damaged 90% of homes and left over 100,000 residents without electricity, water and telephone service. Finally, telecommunications also play a critical role in facilitating the reconstruction process and coordinating the effort of getting returnees displaced by disasters back to their original homes. Other telecommunication applications ranging from remote sensing and global positioning system (GPS) to the Internet and Global Mobile Personal Communications via Satellite (GMPCS), have a critical role to play in tracking approaching hazards, alerting authorities, warning affected populations, coordinating relief operations, assessing damages and mobilizing support for reconstruction [4].

1.3 Challenges

Several challenges have to be faced when dealing with telecommunications, and in general ICT, for disaster prevention and recovery [5]:

- Disasters such as the Indian Ocean tsunami in 2004 have underscored the importance of ICT, including space-based communication tools, for supporting effective disaster reduction practices. This importance has also been acknowledged in international initiatives, such as in the Declaration of Principles and the Geneva Plan of Action of the World Summit on the Information Society; the latter specifically mentions the use of ICT applications in, among other areas, the provision of humanitarian assistance for disaster relief. *It is an ongoing challenge that developing countries have limited access to ICT.*
- Higher priority must be placed on *compiling and institutionalizing disaster risk information at the regional, national and subnational levels*, through detailed disaster loss databases, applications of indicators and indexes, and detailed risk mapping and analysis. Particular efforts are needed to systematically incorporate such information into national programmes to reduce underlying risks and tailor preparedness for responses to potential risks.
- *Investment in early warning systems* and other measures for disaster reduction, including the development of ICT applications tailored to local conditions, yields considerable benefits, particularly when compared to the potential cost of deciding not to invest. In terms of reducing economic losses, early warning and disaster preparedness pay for themselves many times over throughout the life of the warning system. Reducing impacts and losses has long-term benefits for the economy.
- Facilitated by the rapid evolution of ICT, access to suitable communications tools and the rapid growth in bandwidth, global Web-based access to geospatial information and relevant applications is fast becoming a reality across suitable technology infrastructures. Because the Web is an almost universal platform for distributed computing that integrates diverse information systems, it has been possible to overcome the decades-old technical challenges of *interoperability*. The Web has also facilitated the processing of data, thus adding value to the information used in various applications of ICT, including disaster management.

1.4 Outline

In Chapter 2 it is discussed how computer and communication networks have become critical elements of modern society. These network infrastructure systems not only have global reach, they also have impact on virtually every aspect of human endeavor. They have become principal enabling agents in business, industry, government and defense. Major economic sectors, including, energy, transportation, telecommunications, manufacturing, financial services, health care, and education, all depend on a vast array of networks operating on local, national, and global scales. This complex infrastructure system is linked through vast physical and information-based facilities, networks, and assets which have become completely interdependent, and if disrupted, would seriously impact health, safety and security of citizens or effective functioning of governments and industries.

Avoiding failures in these complex systems is a challenge due to their large-scale, nonlinear, and time dependent behavior where mathematical models describing such systems are typically vague or non-existent. Threats arise from reliance on commercial components of unquantified reliability, unsecured legacy software systems, and a lack of understanding of continuously evolving distributed complex networks. As these sectors continue to grow and expand in a distributed nature, the importance that they be able to resist and circumvent disasters similarly increases. The pervasive societal dependence on network infrastructures magnifies the consequences of intrusions, accidents, and failures, and amplifies the critical importance of ensuring disaster tolerant infrastructure systems. There is a clear need to develop new methods for the design of disaster tolerant systems, and for augmenting existing large-scale systems with a disaster tolerant capability, defined as the ability to *continue operations uninterrupted* despite a disaster. The ability to define and establish system requirements in this regard is a specific area of critical need for successful disaster tolerant system development and is the focus of this chapter.

Chapter 3 explores the role of infrastructure in communication network performance during critical events. Communication systems infrastructure is divided among centralized network elements, such as central office buildings and their air conditioners and power plants, and distributed network elements, such as digital loop carrier remote terminals or wireless systems base stations power plants. Because of their importance within the context of the effect of natural disasters on communication networks, Chapter 3 details the main components of conventional communications power plants: rectifiers, batteries and standby diesel generators and discusses mathematical models to calculate power supply availability; i.e., the probability that the power plant is powering the load at any given time. The analysis indicates that in normal operation conditions, an availability of 0.99 999, as requested as a minimum for most communication system operators, can only be achieved with no less than a few hours of battery backup. Although generators contribute to improving availability from that provided by the power grid, they can only improve power supply availability up to 0.9 999. Since air conditioners are powered from the grid tie point only backed up by generators, their availability is at best 0.9 999. Worst availability performance can be expected during disasters.

Next, Chapter 3 details the effects of many critical events on communication networks from 1980s up to the present, including the last significant natural disaster, the February 2010 earthquake in Chile. From this historical description, infrastructure-related issues and, in particular, power supply problems, are identified as a cause of many outages affecting communication networks during disasters. The last part of this Chapter is dedicated to describe commonly used restoration practices when disasters affect infrastructure elements and to discuss alternatives to mitigate the effect of natural disasters on communication systems infrastructure. Three important periods for the analysis are considered: during the disasters, the immediate aftermath, and the long term aftermath. In particular, the use of small local power grids powered by their own set of diverse electric generators, or micro-grids, is identified as a potential suitable solution to power grid issues in all of these three phases when the effects of a disaster is felt on a communication site. It is expected that micro-grids will enable the development of advanced smart power grids which, in turn, will improve overall power supply availability, and, hence, improve communication systems performance to critical events.

In Chapter 4, one introduces and characterizes what is a disaster and which characteristics should a network have to be successfully deployed in such a scenario, so to serve as support to the disaster relief operations. One introduces the Cognitive Radio and the Self Organizing/Autonomic Networks paradigms, and show how they can be combined to enhance a Disaster Relief Network.

The chapter concludes with an overview and discussion about the possible deployment scenarios, focusing on the different types of network topology, like Cellullar, Ad-hoc, Mesh and Satellite, proposing in the end the most balanced topology, which is the mesh based one.

In Chapter 5, one discusses that as emergency communication systems need to be as reliable as possible, the performance of such systems has to be evaluated thoroughly. When creating a scenario for performance evaluation of communication systems, modeling mobility and traffic is an important task. The results of the evaluations strongly depend on the models used.

In this chapter an overview of models available for realistic modeling of public safety scenarios is presented. Detailed models for mobility, traffic, and traffic source distribution have been developed and are described. After this, in two different sample evaluations the impact of the detailed modeling on the network performance evaluation is examined. The specific models allow realistic scenario modeling and show significant impact on the results.

In Chapter 6, we discuss issues related to internetwork operability and multi-system radio resource sharing for supporting disaster relief and emergency communication. Many research efforts have been carried out for investigating heterogeneous emergency systems where the inter-operation of different wireless communications networks has been proposed for efficient emergency communication. Commercial cellular mobile networks, infrastructure-less ad hoc networks and satellite networks play an important role in disaster mitigation and relief. In an emergency communication system, ad hoc mobile networks can be used for establishing self-deployment and auto-configurable emergency communication clusters in which survivors and disaster responders can communicate with each other in the hop-by-hop model. Communication between such clusters can be provided by either satellites or by airships. Further, if mobile terminals are equipped with several radios including WLAN and mobile cellular radios, mobile users can get better communications quality and higher survivability.

Commercial cellular mobile networks become important facilities for providing emergency information services to disaster survivors, as the number of mobile subscribers is very high in most countries. System performance benefits gained by radio resource sharing between 3G mobile networks are also presented, showing the need of future cooperation between mobile operators in case of disasters. Network facilities and mobile terminals should also be equipped with emergency supported functionalities. For further research, system performance evaluation using practical incoming traffic information needs to be carried out. The emergency scenario, in which base stations of 3G mobile networks are destroyed, is also an interesting research topic. Design of mobile terminals and system test beds is needed to improve and validate the proposal of multi-system resource sharing method presented here.

In Chapter 7, the authors discuss how rapid developments in wireless communications systems and the introduction of new technologies such as cognitive radios bring beneficial methodologies and solutions to the problems of public safety and emergency communications, especially related to interoperability issues. Physical layer adaptation and spectrum sensing methodology of cognitive radios are commonly emphasized related to spectral efficiency and frequency domain interoperability of public safety communications. However, wide range of opportunities introduced by the awareness, learning and intelligence features of cognitive radios necessitate the reassessment of opportunities to the public safety and emergency case communications from a more complementary aspect. One application field can be the development of methods that can lead to communicate, locate and reach victims who are stuck in disaster areas, underground (e.g. underground mine explosions) or behind obstacles. Robust and fast detection of the victims' location during emergency cases such as vast area floods and other natural disasters is an important part of public safety communications. Awareness, learning, and intelligence features of cognitive radios can be beneficial for victim and first responder location estimation. Location and environment based services of cognitive radio technology can be employed to detect and locate victims even for cases in which the original core wireless communications network is down.

In this chapter, first the evolution of the cognitive radio systems is discussed along with the technical difficulties that the first responders encounter in the field. The solutions to these problems which are provided by the regulatory organizations, governmental and private institutions are also explained in their historical progress. Second, the general opportunities which are introduced by the cognitive radio technology to the public safety communications are defined shortly. Finally location awareness feature of cognitive radio systems is defined in the context of public safety communications. It is aimed to discuss how communication with survivors can be accomplished and how the estimation of their locations utilizing received signals can be achieved by benefiting from the advantages of cognitive radio technology. Location awareness engine, location sending sensors and related methods are specified along with a Received Signal Strength Indicator (RSSI) based signal detection and power estimation algorithm for location estimation methods.

Networks of mobile robots for rescue operations are described in Chapter 8. The mobile robots and the base station (BS) are connected by a wireless ad hoc network during their integrated operations in disaster areas. Following the explanation of the fundamentals of ad hoc networking, autonomous chain network formations by the robots and their wireless quality of service (QoS) networking are introduced. The chain networks are essential for reconnaissance into distant spaces within collapsed buildings or so, where victims may be trapped. The QoS networking reserves the transmission bandwidth from robots to BS for wideband signal transmissions such as dynamic picture images showing the victims.

2

Systems Engineering Methodology for the Development of Disaster Tolerant Networks

Michael Harper

Science Advisor of the Office of Naval Research, US Navy, USA;
e-mail: harper.michael@gmail.com

2.1 Introduction

Quality of life in modern society relies, in large measure, on the continuous operations of a complex infrastructure. This infrastructure is comprised of physical and information-based facilities, networks, and assets, which if disrupted would seriously impact health, safety and security of citizens or effective functioning of governments and industries. Infrastructure systems include telecommunications, energy, banking and financial, transportation, water, healthcare, government and emergency systems. All of these systems are linked through vast physical and cyber networks which have become completely interdependent. These networks present with a multitude of distributed heterogeneous components so tightly interconnected that any focal point disaster can lead to widespread failure almost instantaneously.

This proliferation of geographically distributed, interconnected, and complex networks throughout both governments and the private sector has increased the vulnerability for cascading failures with widespread consequences. Secure and reliable operation of these systems is fundamental to the economy, national security, and the quality of life of a nation. However, avoiding failures in complex IT application systems is a challenge due to their large-scale, nonlinear, and time dependent behavior where mathematical models describing such systems are typically vague or non-existent. Critical sectors of our society are becoming increasingly dependent upon

N. Marchetti (ed.), Telecommunications in Disaster Areas, 11–51.
© 2010 *River Publishers. All rights reserved.*

highly distributed information systems that operate in unbounded networks, such as the Internet. Threats arise from reliance on commercial components of unquantified reliability, unsecured legacy software systems, and a lack of understanding of continuously evolving distributed complex networks. All of these variables are vulnerable to outside manipulation through network exploitation and outsourcing of system design and development to third party contractors. As these sectors continue to grow and expand in a distributed nature, the importance that they be able to resist and circumvent disasters similarly increases. These factors, combined with recent global events, including natural disasters and terrorist activities, emphasize the need that a disaster event must be considered in the requirements planning and early design phase of system development.

Computer and communication networks have become critical elements of modern society. These network infrastructure systems not only have global reach, they also have impact on virtually every aspect of human endeavor. They have become principal enabling agents in business, industry, government and defense. Major economic sectors, including defense, energy, transportation, telecommunications, manufacturing, financial services, health care, and education, all depend on a vast array of networks operating on local, national, and global scales. This pervasive societal dependency on network infrastructures magnifies the consequences of intrusions, accidents, and failures, and amplifies the critical importance of ensuring disaster tolerant infrastructure systems.

A disaster is an event that can cause system-wide malfunction as a result of one or more failures within a system. Disasters may occur as the result of single or multi-point failure and may occur either simultaneously or sequentially. Disaster tolerance is a superset of fault tolerance in that a disaster may be caused by multiple points of failure in a system that occur very close together in time as well as a single point of failure that escalates into a wide catastrophic system failure [6]. Adequate means to ensure continued system operation in the event of a disaster requires highly reliable and survivable system design of distributed and interdependent systems. It is in this respect that system engineering and the system process emerge as the focal point for the design of disaster tolerant computer and communication systems.

The model for disaster tolerance is different with respect to fault tolerance since it is assumed that failures occur due to massive numbers of individual faults. Disasters may be either natural, such as a flood, or man-made, such as a terrorist event. In either case, the system model is the one of multiple individual system faults that occur nearly simultaneously or close together in time as a series of related events [6]. Fault tolerant system research and development has been evolving for many years out of the even older field of reliability analysis. The typical high-level approach to fault tolerant system design is as follows:

- Determine a model for a system;
- Characterize a failure as a fault model;
- Analyze the behavior of the system with the fault model;
- Redesign the system to function correctly even when the fault is present.

This approach has led to many innovations and positive results in the development of systems that are more fault-tolerant than those developed prior to this period. Unfortunately, most of these successes have been based on the assumption of a single fault being present [6]. This is a valid assumption for modeling systems at burn-in time or those systems near the end of their lifetime since typically a single component will fail causing system failure if it is not tolerant to the fault. While the area of fault tolerance has reached a degree of maturity, disaster tolerant engineering practices are in their infancy.

Disaster tolerance is a superset of fault tolerance methods in that disaster tolerance requires robustness and adaptability under conditions of intrusion, system failure, or accident. For research and analysis purposes in this regard, a failure is defined as a malfunction caused by a fault. The notion that disaster tolerance is identified as a superset of fault tolerance includes the fact that fault tolerance relates to the statistical probability of an accidental fault or combination of faults, not to malicious attack. For example, an analysis of a system may determine that the simultaneous occurrence of three statistically independent faults ($f1$, $f2$, and $f3$) will cause the system to fail. The probability of the three independent faults occurring simultaneously by accident may be extremely small, but an intelligent adversary with knowledge of the system's internals can orchestrate the simultaneous occurrence of these three faults and bring down the system. A fault tolerant system most likely does not address the possibility of the three faults occurring simultaneously, if the probability of occurrence is below a threshold of concern. A disaster tolerant system must take these probabilities into account and provide adequate counter measures to deal with the possibility.

There is a clear need to develop new methods for the design of disaster tolerant systems and for augmenting existing large-scale systems with a disaster tolerant capability. The ability to define and establish system requirements, in this regard, is the area of critical need for successful disaster tolerant system development.

This chapter outlines a methodology for the development of complex disaster tolerant system requirements, specifically, establishing minimum disaster-free operating periods and providing the ability to measure risk through an evaluation of potential loss based on a proposed system design. This methodology will allow a system designer to determine the expected losses from multiple system failures in a proposed system network design structure so that proper quantitative disaster tolerant requirements may be established. Specifically, complex systems are developed with less risk to catastrophic failure by establishing disaster tolerant requirements in the preliminary design stage.

2.2 The Systems Engineering Process

Systems Engineering (SE) is an interdisciplinary approach encompassing the entire set of scientific, technical, and managerial efforts needed to provide a set of life-cycle balanced system solutions that satisfy customer needs. The standard objective in the systems engineering process is defining the minimum essential work products needed to adequately define a system over its life cycle such that all integrated system elements provides at least the minimum needed capabilities with respect to cost, schedule, and risk, in addition to allowing for the potential for evolutionary growth [7]. Systems Engineering focuses on defining customer needs and required functionality early in the development cycle, documenting requirements then proceeding with design synthesis and system validation while keeping under consideration the intended system lifecycle.

The systems engineering approach to developing requirements and design, as outlined in Figure 2.2, is best thought of as a series of activities that together enable a set of "capability needs" to be understood and analyzed in order to develop systems that will meet these stated "capability needs". This approach consists of requirements definition and analysis, functional analysis and design synthesis [8]. The requirement definition and analysis phase determines scoping, operational concepts and required specifications of the system under consideration. The functional analysis phase is a logical and systematic approach that translates system operational and support requirements into specific design requirements. These designs will eventually lead to systems which will be tested to verify that they perform as expected before released for operational use. Requirements are at the heart of good systems engineering. The failure to develop good statements of requirements and the

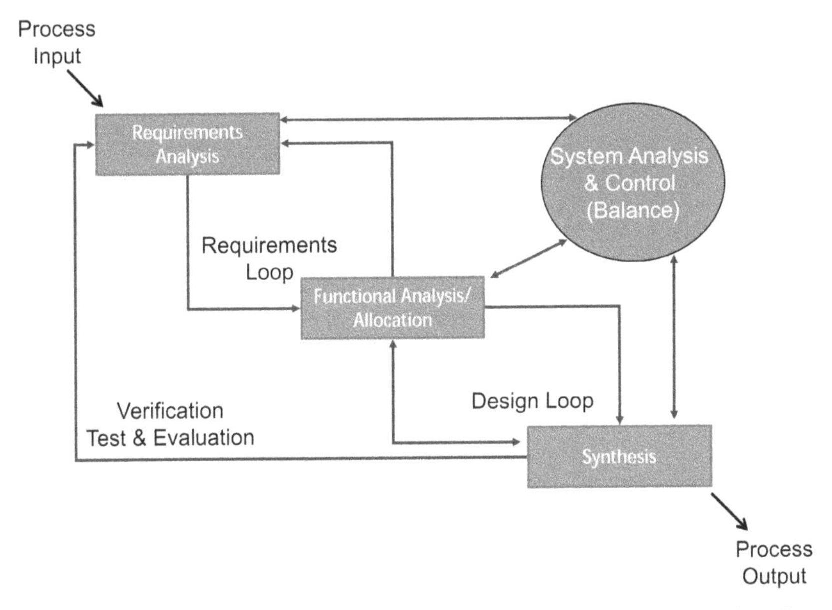

Figure 2.1 System engineering approach to developing requirements and design. *Source: Defense Acquisition Guide, Chapter 4 Systems Engineering*

failure to manage requirements, once developed, is probably the major cause of failures in systems engineering [8].

Systems Engineering starts with analysis of the top-level (system) requirements that are decomposed into lower level (subsystem) requirements, which in turn are decomposed into detailed requirements at the lowest level, this phase is known as functional decomposition. The main purpose of functional decomposition is to break up large or complex system operation into smaller, more manageable parts to facilitate understanding of system operation. The "Systems Engineering Development Process" as illustrated in Figure 2.2, represents the phases of the lifecycle. The left-hand side of the "V" represents the design leg; it starts at the top, proceeds to intermediate level and finally to lowest level.

System fabrication is represented on the right hand leg of the "V" and is a bottom-up process. Components developed during the fabrication process are either manufactured or bought, and are integrated into lower level assemblies. The assemblies are then integrated into subsystems, which are further integrated into configuration items and finally into fully integrated systems [8].

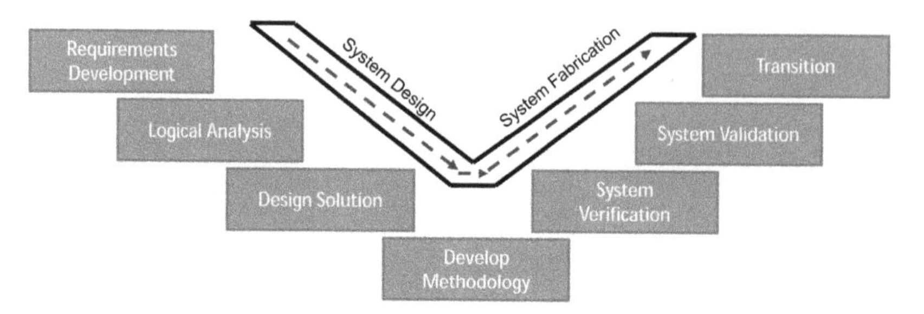

Figure 2.2 Systems engineering development process. *Source: Defense Acquisition Guide, Chapter 4 Systems Engineering*

At each level, the parts, assemblies, and items are tested to verify that they perform as expected until there is an integrated and tested system.

2.3 Chapter Objective

The vulnerability of critical infrastructure to disasters, specifically within the computer and communication systems domain, creates new challenges for systems engineering, as well as for research across related disciplines such as software engineering, computer science and security engineering. In systems engineering, design for disaster tolerance including survivability, reliability and fault tolerance, emerges as a critical research priority.

As previously established, disaster tolerance is a superset of fault tolerance. The model for disaster tolerance is different since it is assumed that failures occur due to massive numbers of individual faults. Disasters may be either natural, such as a flood, or man-made, such as a terrorist event. In either case, the system model is one of multiple individual system faults that occur nearly simultaneously or close together in time as a series of related events [6]. A nave way to provide disaster tolerance in a system is to utilize redundancy with redundant components geo-located in different areas; however, this approach has two serious consequences:

1. Communication between the redundant systems becomes a critical link and redundant communication channels may also be required.
2. Some systems are so large that it is impractical to replicate them (for example, the United States electric power grid).

There is a distinct difference between disaster tolerance and disaster recovery. Disaster recovery is the ability to *resume* operations after a disaster, whereas disaster tolerance is the ability to *continue operations uninterrupted* despite a disaster. Developing a system that adapts to preserve essential services involves identifying the mission-critical applications and the availability requirements to provide necessary support for operational success [9–11].

To better understand the evolution of disaster tolerant research, it is first necessary to understand the motivation to design systems with disaster tolerant capabilities, which is a product of the large scale, complex systems being designed today. The enormously large scale of many critical infrastructure systems, particularly in software, communications and computing systems, and military systems, including the large cost of down time, both financially and militarily, necessitates an approach for independent subsystem modeling and design in order for the problem to become feasible. One of the biggest obstacles, which has hindered research in the area of large scale, complex systems, is the inability to develop appropriate models at this scale, particularly integrated software, hardware, and networked systems, in a tractable amount of time with a suitable degree of detail, so that system behavior can be obtained both in normal operating mode and in the presence of a disaster [11].

2.4 Evolution of Disaster Tolerant Research

As mentioned above, fault tolerant design means providing a system with the ability to operate, perhaps at a degraded but acceptable level, in the presence of a fault [11–13]. The standard fault-error failure model is outlined in Figure 2.3. A fault is an imperfection or deficiency in system operation. However, a program may operate as intended until some triggering input condition is encountered. Once triggered, some of the computed values will deviate from the design intent (an error). However, the deviation may not be large enough, or persist long enough, so the system may recover naturally from the "glitch" in subsequent computations, also known as 'self-healing'. Alternatively, explicit design features such as diversity or safety kernels can be used to detect such deviations and either recover the correct value (error recovery) or override the value with a safe alternative (fail-safety) [12–14].

Transition among states in this model depends on (1) if fault tolerance has been designed in, or (2) nature of the application (i.e 'grace' time, defined as the rate at which transition to a safe operating mode occurs, self-healing). The probability of a transition depends on the likely number of faults, the

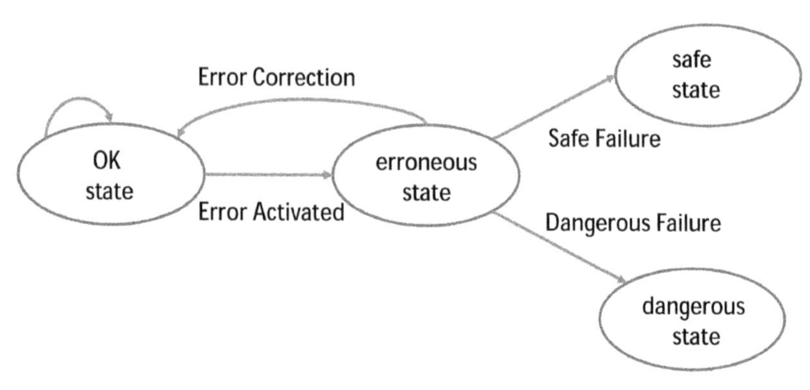

Figure 2.3 Model of system failure behavior

fault size and its impact to the system at-large, and the impact towards the intended use of the application [12]. The transition between either a safe state or a dangerous state will depend on the elements of a fail-safe design and what hazards may exist.

2.5 Limits in Fault Tolerance Effectiveness

The imperfections of fault tolerance, such as the lack of fault tolerance coverage, referring to the notion that all system elements may not be protected through error correction and be protected from entering an erroneous state, constitute a severe limitation to the increase in system wide, dispersed attacks. These limitations are due either to (1) design faults affecting the fault tolerance mechanisms with respect to the fault assumptions during the design, the consequence of which is a lack of error and fault handling coverage, or (2) to fault assumptions which differ from the faults really occurring in operation, resulting in a lack of fault assumption coverage [15,16]. This can be attributed to either failure mode coverage, failed component(s) not behaving as assumed or a lack of failure independence coverage as outlined in Figure 2.4.

2.6 Impact of Application Downtime

The Internet is an example of an unbounded environment with many client-server networked applications [17]. Users of the Internet exist within a network-centric environment operating in many different administrative do-

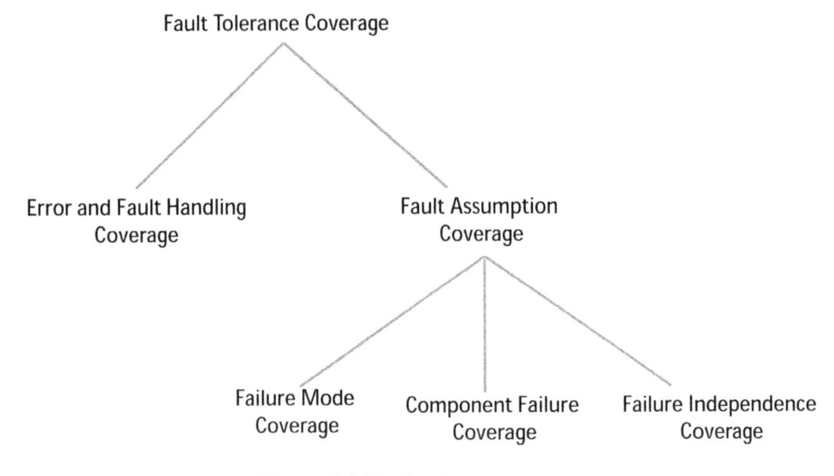

Figure 2.4 Fault tolerant coverage

mains. Many business-to-business Web-based e-commerce applications depend on conventions within a specific industry segment for interoperability. In the military setting, network-centric interoperability has been adopted through initiatives such as the United States Department of Navy's FORCENet architecture which integrates warriors, sensors, networks, command and control, platforms, and weapons into a networked and distributed combat force [18]. The ability to acquire and maintain information superiority demands intense requirements of availability and disaster tolerance.

The capability to deliver essential services in a constant and continuous manner must be sustained even if a significant portion of the system is incapacitated. This capability should not be dependent on the survival of a specific information resource, node or communication link. In a wartime environment, essential services might be those required to maintain technical superiority and essential properties may include integrity, confidentiality, and a level of performance sufficient to deliver results in less than one decision cycle of the enemy. Similarly, in the public sector, an IT application system maintaining financial information has the requirements to maintain integrity, confidentiality, and availability of essential information and financial services, even if particular nodes or communication links are incapacitated due to a debilitating event [19].

A methodology and model to establish disaster tolerant requirements and goals is an essential design related activity. Without understanding what is required, undesirable results may occur, including the potential that the product will not satisfy the customer or resources will be wasted (i.e., overdesign). While defining a system with disaster tolerant capabilities gives direction to subsequent efforts, developing the specific disaster tolerant goals and requirements scopes the magnitude of those efforts. The process of decomposing the activities leading to disaster tolerant requirements is the requirements engineering process. This includes proper formulation of requirements engineering development phases within the systems engineering process where many potentials problems may exist. There are many issues associated with requirements engineering, including problems in defining the system scope, problems in fostering understanding among the different communities affected by the development of a given system, and problems in dealing with the volatile nature of requirements [20]. These problems may lead to development of poor requirements, resulting in development of a system that is judged unsatisfactory or unacceptable, has high maintenance costs, undergoes frequent changes and potentially resulting in the cancellation of the total system development. The impact of poor requirements engineering in areas such as weapon systems development has the potential for loss of life.

2.7 Systems Development

Referring back to the discussion of the systems engineering development process and illustrated in Figure 2.2, it is important to outline these aspects as they relate to the notion of incorporating disaster events within this process. Specific attention is directed at the requirements development phase as is the focus of this chapter.

2.7.1 Requirements Development

A requirement is the minimum level of performance acceptable to, or expected by, the customer. In contrast to a requirement, a goal is usually the level of metric being established, such as reliability, which is greater than that required. Early in the development of a new system, goals may be used as a starting point to challenge the development to make use of innovative and new design approaches. As development progresses, these goals evolve into firm requirements. Firm requirements should meet or exceed the customer's needs and expectations [21–23]. A requirement is a "function or characteristic of a system that is necessary; the quantifiable and verifiable behaviors that a system must possess; the constraints that a system must work within to satisfy

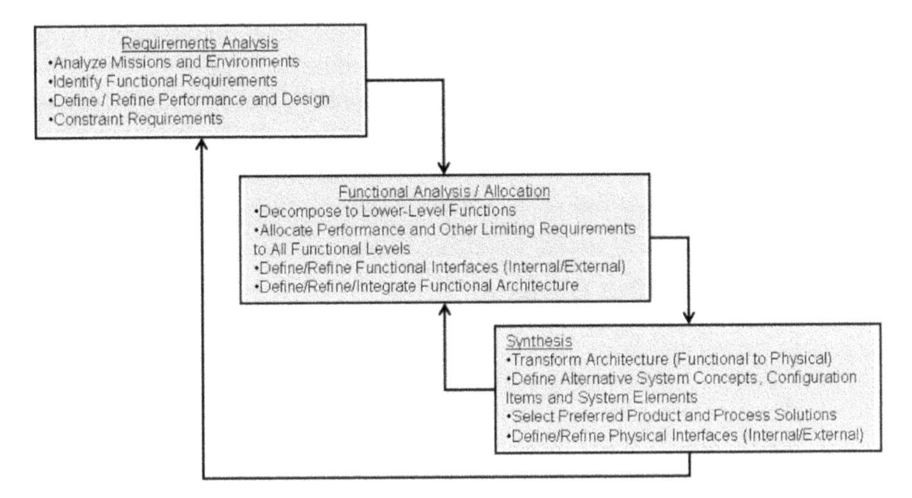

Figure 2.5 Requirements decomposition process

an organization's objectives and solve a set of "problems" [24]. Similarly, the Institute of Electrical and Electronics Engineers (IEEE) defines a "requirement" in the following manner: (1) a condition or capability needed by a user to solve a problem or achieve an objective; (2) a condition or capability that must be met or possessed by a system or system component to satisfy a contract, standard, specification, or other formally imposed documents; (3) a documented representation of a condition or capability as in (1) or (2) [24].

To fully understand requirements at all levels of a system requires top-level requirements be "decomposed" into lower level requirements. For example, a user may require that a given system "be capable of withstanding a targeted attack." The user might state certain performance requirements associated with tolerating that attack; for example, it may be required the system be tolerant of withstanding an attack within certain time parameters. To fully comprehend this requirement and its implications may demand that the "attack tolerance" function be decomposed into lower level requirements, i.e., consider what makes a system tolerable of an attack. Figure 2.5 illustrates the iterative process to decompose requirements into a design solution.

As the top-level system requirements are decomposed into lower-level requirements and further analysis of the performance is conducted, more requirements may be discovered that were not originally stated or the original requirement is determined to be more complex than originally thought

[21, 25]. This requires that the sub-functions be arranged into a logical order which best meet the user's needs. To incorporate the consideration of disaster events within the systems engineering development process, specific disaster tolerant requirements must be developed. In this regard, disaster tolerant requirements must guarantee not only the required minimum availability but must also reduce the risk of failure below a maximum acceptable level.

2.8 Addressing the Disaster Tolerant Problem

The risks associated with the reliance on interconnected computer systems are substantial. By launching attacks across a span of communications systems and computers, attackers can effectively disguise their identity, location, and intent, thereby making them difficult and time consuming to trace. Such attacks could severely disrupt computer supported operations, compromise the confidentiality of sensitive information, and diminish the integrity of critical data. It is necessary to first address the various types of failures that may be exhibited in a complex system.

2.8.1 Types of Failures

When viewing from a system's perspective, there are many types of failures. A critical failure is considered within the context of this chapter to be a critical event which leads to a system halt or degeneration of the required function below a minimum acceptable level and requires immediate intervention for repair. However, it is critical that the myriad types of failures are discussed, as their understanding within the disaster tolerance research area is imperative.

Cascade Failure

Cascades of failure are those that can spread quickly from one network to another through "input" and "mutual" interdependencies. Dynamic movement of information, power, and data creates vulnerabilities that can be exploited. The loss of a critical node can result in system-wide cascading failure.

Escalating Failure

The failure in one networked infrastructure can exacerbate a failure in another network. This failure is typically due to "shared" or "exclusivity" interdependencies. For example: an attack against transportation network would slow repair of an electricity failure.

Common Cause Failure

This failure is due to a single disruption that directly impacts two or more networks. This failure is typically due to geographical "co-location."

Infrastructure Failures

In today's complex infrastructure systems, failures are not limited to a single network. They spread across networks due to a complex interplay of interdependencies. These interdependencies are both tightly coupled, where connections rapidly spread a failure to other systems and are non-linear, in which feedback loops magnify the failure impact. Network interdependencies can be classified into five categories as follows:

1. Input: information delivered by one network is used by another;
2. Mutual: networks that serve as inputs for each other;
3. Co-location: different networks located in the same geographical area;
4. Shared: a network which shares physical components, transport, or facilities;
5. Exclusive: a network that can only support one or few outputs.

Current infrastructure designs maintain various fault tolerant methods for addressing these situations, including load redistribution and high-load nodes. In this respect load redistribution refers to situations where loads carried by each node on the network are dynamically redistributed. If a network node is lost, due to disaster, the load that node carries is rapidly distributed to the other nodes on the network.

Furthermore, if a high-load node is removed from the network, the loads it carries are redistributed to other nodes on the network. The increased capacity on remaining nodes provides less capable nodes to potentially exceed their capacity. To protect these nodes from damage, many networks will automatically force the overloaded node to fail-over (shut down), or the node will simply result in increased congestion causing the overloaded node to become inefficient. The result is series of shut-downs that "cascade" through the network as the excess load is pushed to the next available node with end result of total network failure [26].

2.9 Disaster Tolerant Systems Engineering

Fault tolerance of a system is essential to ensure continued operation and provide necessary system services despite the failure of components. The goal of a fault tolerant system can be defined as the specified degree of resiliency in a system, subject to minimizing overhead costs such as duplicate resources, communication, and time overheads [27].

When disrupted, an information system must be adjusted to ensure continued provision of the information services on which the infrastructure depends. Adjustment will involve reconfiguration. To be reconfigured, an information system must of course be reconfigurable. System reconfigurability can occur at many levels, including operating parameters, module implementations, code location, replacement of physical device, etc. An understanding of the failure rates inherent in these levels requires incorporating systems reliability.

It must be emphasized that the fault tolerant research has not resulted in a single solution or design methodology to all fault tolerance problems. There are a myriad of techniques within the fault tolerant arena which are categorized as follows.

2.9.1 Massive Redundancy

Massive Redundancy is the concept of providing multiple identical instances of the same system and switching to one of the remaining instances in case of a failure. In this manner, diversity can be built in which provides multiple different implementations of the same specification, and using them like replicated systems to cope with errors in a specific implementation [28].

2.9.2 Full System Backup

Full System Backup refers to making copies of data so that additional copies may be used to restore the original after a data loss event. Backups are useful primarily for two purposes. The first is to restore a state following a disaster and the second is to restore small numbers of files after they have been accidentally deleted or corrupted. Data storage requirements are considerable in a full system backup methodology as at least one copy of all data is stored at all times.

2.9.3 Duplication

Duplication is the classic hardware redundant method for error detection and simply compares the results of two copies of a circuit. It is fast, but cannot correct an error [29]. This method has been extended to incorporate duplicated elements which work in parallel. A typical example is a complex computer system that has duplicated nodes, so that should one node fail, another is ready to carry on its work.

2.9.4 Triple Modular Redundancy

The Triple Modular Redundancy (TMR) technique involves the use of three identical systems performing the same task in parallel with corresponding outputs being compared through a majority voter circuit [30]. The voting circuit can determine which replication is in error when a two-to-one vote is observed. In this respect, the voting circuit has a high probability of outputting the correct result, and discarding the erroneous version. After this, the internal state of the erroneous replication is assumed to be different from that of the other two, and the voting circuit can switch to a simple duplication mode with one remaining backup system . This model can be applied to any larger number of replications [31].

2.9.5 Quad Systems

Another redundancy technique known as "quadded logic" has been developed which uses quadruplication in each stage of a network. Failure restoration is accomplished by mixing the four output signals pairwise at the inputs of the next stage. Thus failure is corrected just downstream of the stage at which a failure occurred, with the help of correct signals from the neighboring gates [32]. Quad systems suffer from the law of diminishing returns on reliability in that the inherent failure rate rises in proportion to the increase in components and complexity [33, 34].

2.9.6 Spare Switching

The combination of N-modular redundancy (NMR) and standby sparing, whereby replicate components or systems are available for failover, provides a redundancy technique for protecting the portions of a fault-tolerant system whose continuous real-time operation is essential [35]. This technique, known as hybrid redundancy, uses $N + Sp$ identical modules. N of these are connected to a majority voter to form an NMR core. The remaining Sp are used as standby spares. Disagreement detectors instruct a switch to replace with a standby spare any of the N modules that disagrees with the majority consensus. Since the switch and disagreement detector, as well as the modules, must function properly for the system to perform its designed task, the overall system reliability depends on the disagreement detector, switch, and component reliabilities. Therefore, the overall reliability is a function of switch reliability. A highly reliable, thus simple, switch is desirable [35, 36].

2.9.7 Substitution

Substitution methods for fault tolerance have been proposed for systems addressing fault tolerance for wide area replication of distributed systems. Constructing logical machines out of collections of physical machines is a well known technique for improving the robustness and fault tolerance of distributed systems. In this regard, the physical machine at each site implements a logical machine by running a local state machine replication protocol, and a wide-area replication protocol among the logical machines.

2.9.8 Transmission Codes

In transmission of information or data, there is often the need for processes enabling transmission errors to be detected and attempts to correct the errors. Transmission codes is a method for transmitting information while having the ability to correct individual errors in a burst of errors [37]. Examples of transmission coding includes the calculation of coefficients of code words by obtaining polynomials from the factorization over a field of elements. On transmission, a product of plurality of polynomials is obtained where the product is divided by another polynomial to obtain coefficients. In reception, a reciprocal of the polynomial is used and coefficients are calculated. By summing the most significant coefficients of the polynomials, corrected bits of the message are obtained. This method can quickly become highly complicated and impracticable as certain coding techniques may require a fairly considerable number of redundant bits [38].

2.9.9 Arithmetic Codes

Arithmetic codes falls under the classification of error-detecting codes which can protect large arithmetic expressions involving addition and multiplication, or more specifically, circuits that implement such expressions [39]. Research to detect errors in arithmetic expressions has shown to be poor, due to error masking in multipliers.

2.9.10 Reconfiguration

Reconfiguration is discussed in the literature as a method of replacement for "dynamic redundancy" in regards to the use of interconnection networks for fault tolerance. The goal of reconfiguration for fault tolerance is not the prevention of failure, but rather the manipulation of failure. In this sense, reconfiguration is the ability of a system to alter the active interconnection among modules [40,41]. This method introduces a great deal of complexity as it places extra demands on architecture robustness, including communication within the system, both among the elements in the subset and among the subsets of the entire system [42].

2.10 Examination of Fault Tolerant Methods

Within the context of discussion for this chapter, systems that require continuity of operation within the redundancy domain are examined. First, the full system backup is examined. Figure 2.6 shows a representation of a full backup system. The active system is used for normal operation. If the active system is unable to perform, the backup system can be used. For disaster tolerance, these systems are typically geographically separated. Five classes of disruptive faults are possible. F_1 represents a communication fault between the user(s) and the systems. For this fault the systems and their data would not be corrupted and the systems would be in full operational mode.

As previously mentioned, a failure is defined as a malfunction caused by a fault. For this fault there would be no system malfunction. The magnitude of the communication malfunction can be minimized to a particular user by duplicating the communications path, as shown in Figure 2.7.

If a fault occurs at the user site, that site would become inactive. The remainder of the system would remain operational. F_2 would result in a communication disruption to the active unit. Similarly, F_3 would result in a communication disruption to the backup unit. This would produce no failures in the active or backup units themselves, but the communication link would be lost. These faults could be addressed by duplicating the communications links and geographically separating them, as was done for fault F_1. For F_4, and F_5 as the stand-by system the result would be failures in the active unit and the system could be corrupted. Any of these faults occurring individually would not result in total system failure, thus system data would not be lost or corrupted. However, while the system would remain operational, it would do so in a more vulnerable state.

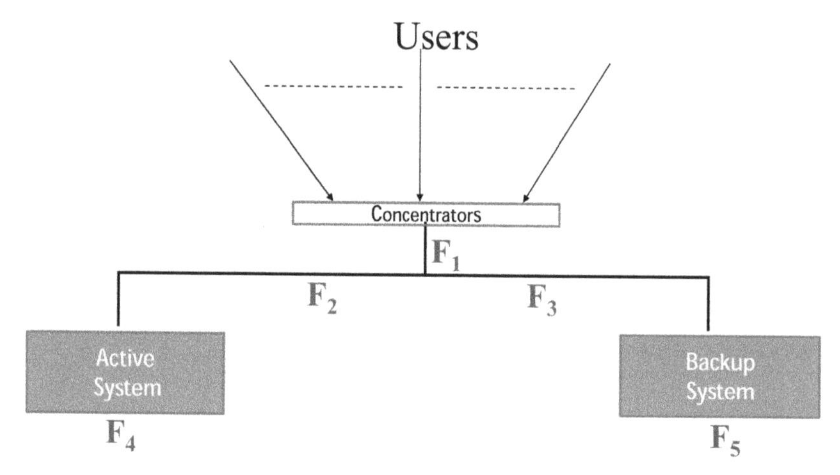

Figure 2.6 Full system backup

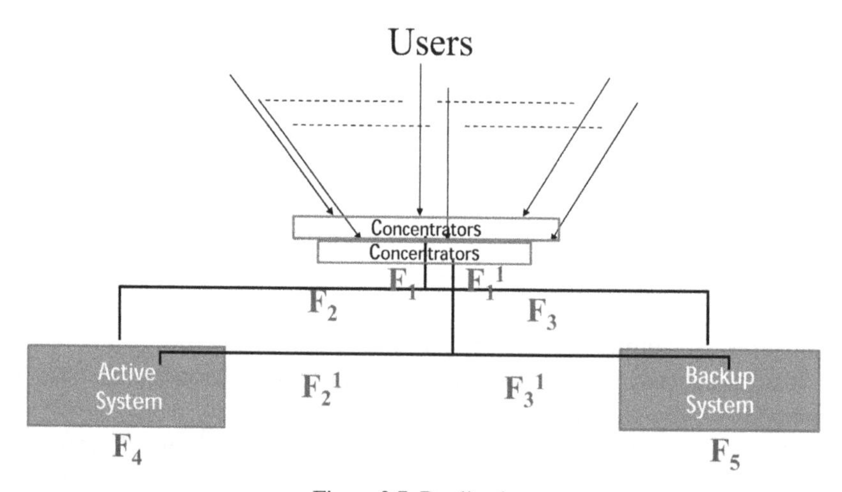

Figure 2.7 Duplication

Unfortunately, there are a number of problems with this system configuration:

1. If a fault occurs, it must be detected and identified to determine which system should be used for continuing operation.
2. The fault needs to be isolated to some replaceable component.

3. The faulty component must be replaced before another fault occurs.
4. For applications such as real time control systems where human life is at risk, items 1–3 must be accomplished without disruption of system operation.

Before considering these points in detail with respect to disaster tolerance, some techniques are addressed which have been used to achieve fault tolerance in various systems over the past 50 years. Extensive research has been performed in the fault tolerant area over the past half of a century, but this research is based on system characteristics and assumptions which are not the same for disaster tolerance. It is an objective of the research within the context of this chapter to determine what can be built upon from this earlier fault tolerant research and fill in the gaps so that effective disaster tolerant requirements may be established early in the system design phase.

Various techniques are applied, based on different requirements, operational considerations, cost and performance constraints, fault models, and threats. Central to this research are the differences between fault tolerance and disaster tolerance. These differences include: threats to the system, fault assumptions, fault models, operational requirements, and cost.

As the basic assumptions are established for disaster tolerant system development, massive redundancy techniques, including triple modular redundancy and quadded system redundancy can theoretically be used. The major drawback for both of these areas is cost and potential performance degradation. The cost quickly becomes prohibitive as you increase the number of replicate systems required to be maintained in the event failover becomes necessary. In this respect, quadded systems are traditionally implemented for only the most critical missions such as applications within the nuclear power field and certain defense system applications. In addition, disaster tolerance must consider long communication links, and catastrophic conditions that must include natural disasters and terrorism that contribute to performance degradation.

Research has been performed in the disaster tolerant area recently, considering both technical and economic problem areas. Much of this work has built on earlier work of Szygenda, Nair, Thornton and the work of others over the past 50 years [6, 7, 13, 37, 42, 43].

A number of specific differences are considered for disaster tolerant systems engineering, including differences in fault assumptions and models, specifically, that fault tolerance is based on the assumption that a single fault will occur and in many techniques it is also assumed that the fault is confined to the failure of an individual component (for example some coding techniques). A second important difference is the cause of the faults; disaster tolerant systems will take into account system wide, catastrophic failures. With respect to fault assumptions and models for fault tolerance,

1. the failure caused by the fault can be masked with techniques such as system reconfiguration;
2. the fault can be isolated, and in some cases repaired before another fault; can occur (the single fault assumption).

For disaster tolerant systems:

1. the single fault assumption is not valid;
2. the nature of the fault can be global;
3. the cause of the fault can be intentional (e.g. caused by a terroristic attack);
4. the scope of the fault can be economically disastrous.
5. In addition to highly critical missions and life threatening situations, threats to our way of life such as cyber related attacks on secure networks designed to maintain and manage personal information online, must also be considered.

Referring back to Figure 2.6, massive redundancy is considered to address the problems specified above, for both fault tolerance and disaster tolerance. It is essential that the technique of Duplication and Match (D&M) be considered. A very notable example of D&M is the AT&T #1 Electronic Switching System that was first put into operation in 1965 and is still being used today.

Consider the pictorial representation, Figure 2.8, of a D&M system. In this configuration, identical systems are used, as was also the case in Figure 2.8. The difference is that the systems are synchronized and the output is compared to determine if they are operating identically and, hence, (based on the single fault assumption) operating correctly. This matching can be performed at any subunit level. The matching components are also duplicated, to address the situation of a possible fault in the matcher.

The location of the matcher can be at the system level, as shown, at a subsystem level, or even at a component level. There is an obvious increase in cost as the number of matchers' increases. However, more matchers at a lower level can help with fault isolation, which is part of the problem indicated for a backup system.

While D&M systems can determine a fault occurrence (a mismatch) they have to resort to interrupt diagnostics based on system configuration to de-

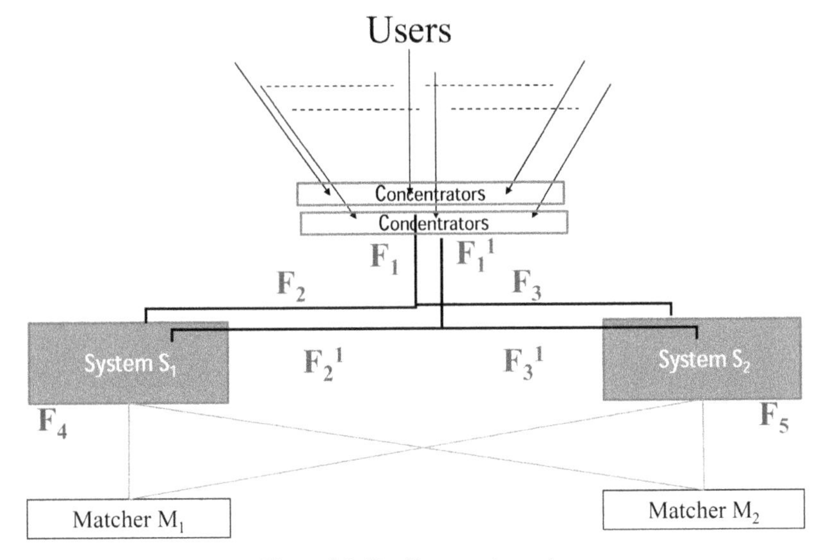

Figure 2.8 Duplicate and match

termine the faulty unit, if for example, matching capability has not been designed into the system architecture. This requires extensive software and interruptions in performance to diagnose the hardware.

After determination of a fault and isolation of the faulty unit, it is removed from the configuration and a working simplex system, defined as one single system in operation with no redundant capability, is put into operation. The simplex system is then capable of leverage processing cycles from the operational system to run software diagnostics and isolate the faulty replaceable component. As long as the replaceable faulty unit is replaced before the Mean-Time-To-Failure (MTTF) of the operational system, the configuration can be re-synchronized and brought back to a D&M system. If the faulty unit cannot be replaced before the MTTF, no redundant capability exists and the probability will increase, based on the failure rate of the components towards total system failure.

This configuration addresses the second problem indicated for the back-up system, being that a fault needs to be isolated to some replaceable component for full system backup. However, a new problem must now be considered. Most D&M systems have been designed to operate in close proximity to one another. This handled the problem due to the single fault as-

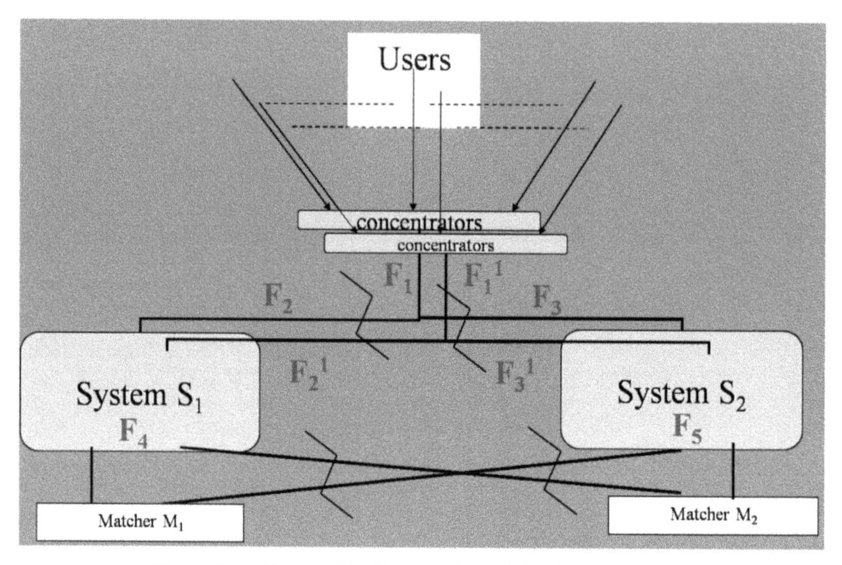

Figure 2.9 Geographical separation of duplicate & match

sumption and the ability to locate and repair the faulty unit and re-synchronize the systems.

For disaster tolerance, one must consider the possibility that the entire system, the building containing the system, or even a geographic area (in case of biological, chemical, or nuclear disaster) could be corrupted. Hence, geographical separation of the duplicated system, as shown in Figure 2.9, must be considered.

A new problem of synchronization is now introduced by the distance separation of the systems, together with the possibility of not being able to repair the faulty system. It is easy to see, based on fault tolerant research, that this configuration would address the backup problems in a disaster tolerant system, as long as the faulty system is repaired before the MTTF of the operational system.

However, this approach has a number of limitations:

1. The cost of hardware is more than two-fold.
2. Considerable hardware communications cost due to distance separation.
3. The cost of software needed for fault identification and isolation is large.
4. The interruption of system performance is needed to determine an operation simplex system. If the system is complex there is a possibility

that the diagnostic tests are not complete and the faulty system cannot be identified or, at least it may take considerable time for isolation.
5. The system must be interrupted to isolate the faulty unit.
6. The previous systems have not been designed to operate as geographically separated by large distances and/or in a network environment.

Other massive redundancy techniques have been developed to address problems 3, 4 and 5 above; namely Triple Modular redundancy (TMR), (the NASA Jet Propulsion Laboratory Self Test and Repair Computer is an early example of such a system), and Quadded Systems (QS) as used for nuclear reactor control. These systems accomplish this by vastly increased hardware costs, some estimates of increased hardware costs for QS is 8 to 10 times that of a duplicate and match system. Before considering these systems we will consider problem 6, which also covers problem 2.

TMR is a fault tolerant form of N-modular redundancy, in which three systems perform a process and that result is processed by a voting system to produce a single output. If any one of the three systems fails, the other two systems can correct and mask the fault. If the voter fails then the complete system will fail. However, in a good TMR system the voter is much more reliable than the other TMR components. Alternatively, if there is another stage of TMR logic following the current one, then three voters can be used - one for each copy of the next stage of logic.

Figure 2.10 shows a system with complete duplication. It should be noted that the duplicate systems could be placed in any node, depending on various considerations, such as threat and distance.

It can be seen that this approach would work in a network environment, but the cost of additional hardware, communications links, software, and synchronization would increase drastically. This configuration can be envisioned as two layers. One layer containing $S_{1,1}$, $S_{2,1}$, $S_{3,1}$, $S_{4,1}$, and layer two containing $S_{1,2}$, $S_{2,2}$, $S_{3,2}$, $S_{4,2}$.

Based on this analysis, it would appear that a D&M configuration could work for disaster tolerance with some limitations and substantial cost and complexity. TMR addresses the problem of real time systems and software needed in D&M systems. This trade off is accomplished with increased hardware, space and power.

A simple TMR configuration, as shown in Figure 2.11, is a simple voting system, i.e. the best 2 out of 3 wins. This system can be used for real time applications since there is no interruption or degradation in performance due to faults. The fault could be massive, as long as it is restricted to one

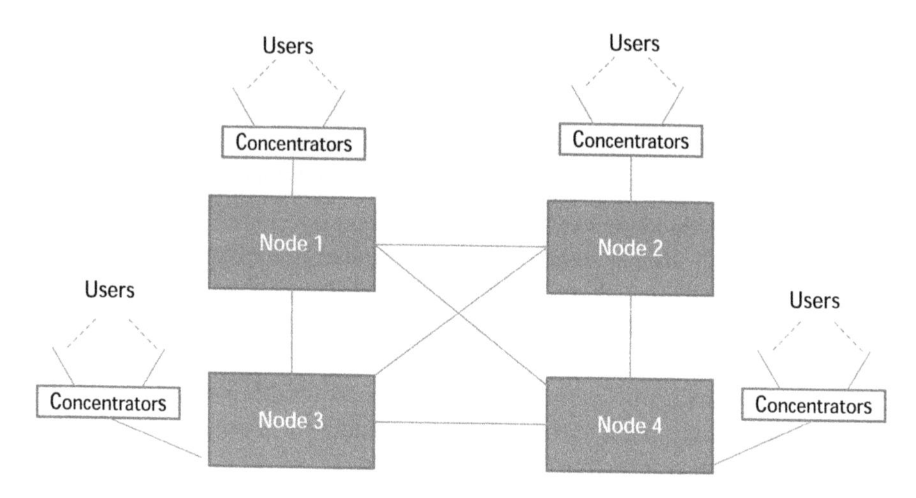

Figure 2.10 Four-node network with complete duplication

Table 2.1 System redundancy vs fault occurrence

	Single Fault	Double Fault	Triple Fault	Quad Fault
Simplex System	No System	No System	No System	No System
Duplex System	Simplex	No System	No System	No System
TMR System	Duplex	Simplex	No System	No System
Quad System	Triplex	Duplex	Simplex	No System

system (site). $G1$ and $G2$ depict geographic separation. Little software is needed for this technique. In a geographically separated system there would be additional cost for communication links and additional synchronization problems.

Table 2.1 illustrates the relationship of redundancy against single and simultaneous fault occurrences. As redundancy increases from a simplex system towards a quadded system, the potential to handle fault similarly increases. However, systems beyond Quad are rarely used because of their cost and the added failure potential due to the increased complexity and number of components. These massively redundant configurations are fault tolerant, given that any fault, regardless of severity, is restricted to a particular node. An added advantage of massively redundant systems is that they are gracefully degradable. That is, if one system fails the other system(s) remain operable.

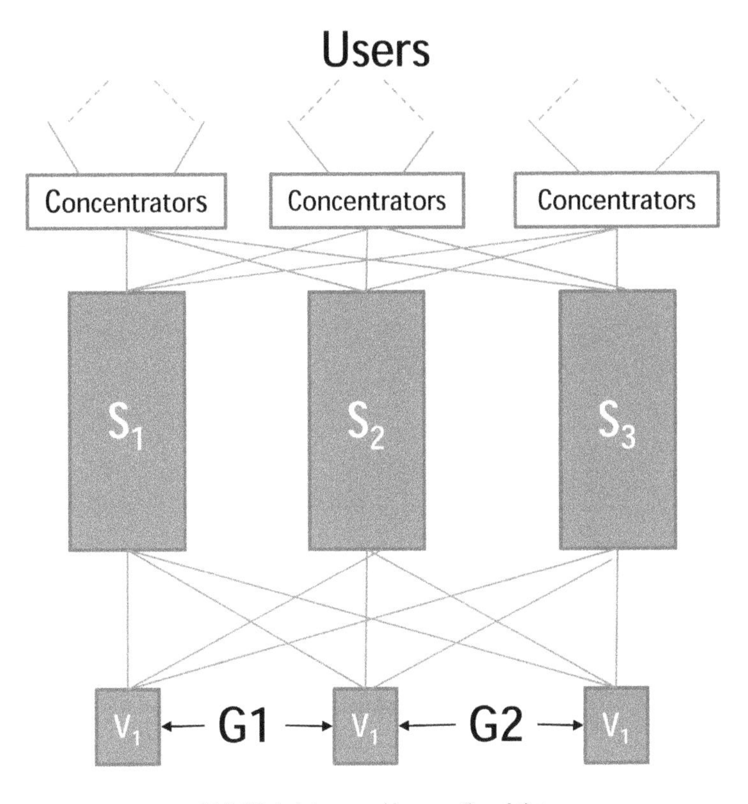

G1 & G2 depict geographic separation of sites.

Figure 2.11 Triple modular redundancy

2.11 Preliminary Design and System Level Disaster Tolerant Requirements

Traditionally, systems development addresses a multitude of factors, including, but not limited to; environment, intended usage, system safety, system reliability, maintainability, and availability. However, a disaster event so far has not been included within the systems engineering design phase. A methodology to include disaster events in the systems engineering process, to ensure the capability of disaster tolerance is necessary. To appropriately capture disaster tolerant requirements, capabilities from a system level must be understood. Figure 2.12 outlines the traditional approach to systems design with the incorporation of disaster events that must be considered in the devel-

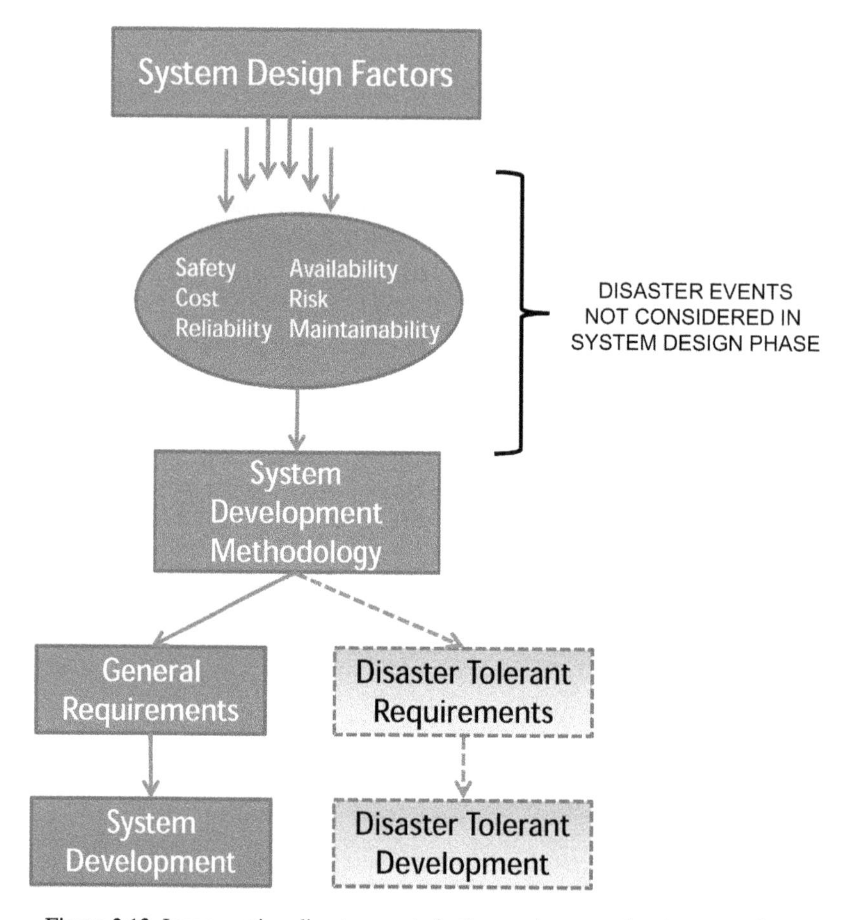

Figure 2.12 Incorporating disaster events in the requirements development phase

opment phase in order to allow a design that will incorporate the capability of tolerating a disaster event.

2.12 Defining Quantitative Disaster Tolerant Requirements

The first step towards incorporation of disaster events in the requirements development phase is to define what comprises quantitative disaster tolerant requirements. Through discussion with subject matter experts in both the fault tolerant and disaster tolerant community, including information security system experts and system designers, it has been determined that a set of system level quantitative disaster tolerant requirements should contain three basic components: (i) a specific minimum availability target; (ii) a specified maximum acceptable level of the expected losses from failures; (iii) a requirement for minimum total losses.

The input data for setting basic disaster tolerant requirements at a system level which limit the expected losses from failures and deliver a required minimum availability target are: (i) the distribution of the downtimes due to a critical failures, (ii) an average cost of the lost production per unit downtime, (iii) distribution of the cost of intervention for repair (includes the cost of locating the failure, the cost of mobilization of resources, the cost of intervention and the cost of repair/replacement).

2.13 Establishing Disaster Tolerance Requirements

A proper formulation of a methodology that provides system designers a quantitative approach to establish disaster tolerant requirements must begin by addressing the a specific Statement of Need. This is specifically defined in the context of this research as the need for a methodology that provides systems designers a means to evaluate a proposed system design based on an evaluation of the potential loss from multiple failures as a result of a disaster event. This will allow the system designer to quantitatively set disaster tolerant requirements based on an understanding of the cost of risk from critical failures due to a disaster event.

Traditional approaches to development of fault tolerant systems have incorporated the premise of identifying and reducing the risk of failure as a result of a fault. These methods have provided numerous advances in secure system design, however many have been based on system failure caused by a single failure scenario. In the very common case where the system can fail due to multiple failures scenarios, these traditional approaches reveal major weaknesses. Often, each individual risk corresponding to the separate failure scenarios is in a low risk region (and therefore perceived as acceptable) of a risk matrix. However, this creates a dangerous perception of security. In many cases, the aggregated risk from all failure scenarios cannot be tolerated without advances to the traditional fault tolerant design model.

Reducing each individual risk below the maximum tolerable level does not necessarily reduce the aggregated (average) risk. A large aggregated risk from multiple failure scenarios, each characterized by risk below the tolerable level can be just as damaging as a large risk resulting from a single critical failure scenario.

What is necessary to know is not the average value of the losses, but the probability that the losses will exceed a maximum acceptable quantity or a critical threshold value. The point of interest for disaster tolerant systems is large losses characterized by small probability of occurring and further defining what is an acceptable cost of risk is to the end user. It is in this regard that the development of disaster tolerant systems must begin with requirements that incorporate the level of risk which limits the cost of failure resulting from a disaster event.

The cost of failure from a disaster can be expressed in monetary units, number of fatalities, lost time, volume of lost production, number of lost customers, or amount of lost sales. In order to minimize the cost of failure resulting from a disaster, a disaster tolerant system must be designed with a low risk of failure. There are three methods to achieve a low risk of failure:

1. Preventive Measures: reduces the likelihood of failure;
2. Protective Measures: reduces the consequences from failure;
3. Combinatorial Measures: includes preventive and protective measures.

These are specific measures that must be addressed during system design so that risk of failure is minimized throughout the lifecycle of the system.

2.14 Risk of Failure Determination

Conditional probability is defined as the probability of some event A, given the occurrence of some other event B. Further, probability theory states that a conditional probability distribution of the event, B, given that A has occurred is the probability distribution of B when A is known to be a particular value. Over a specific time interval, the conditional distribution of B is its distribution conditional on all information available during that specific time interval [44].

Alternatively, unconditional probability is defined as the probability that an event, A, will occur, not contingent on any prior or related results. Unconditional probability is defined as the independent chance that a single outcome results from a sample of possible outcomes. The unconditional probability of a certain event, A, can be determined through summing the outcomes of the event by the total number of possible outcomes.

In this regard, we have the concepts of potential loss and conditional loss as random variables applied to both non-repairable and repairable systems [45]. Conditional loss is defined as the quantity a *loss given failure* as a conditional quantity as it observes the law of conditional probability, that is, an event B occurring when it is known that some event A has already occurred, i.e. in our case, given the fact that a specific failure has occurred [44, 45]. In this regard, the conditional distribution can be used to determine the probability that given failure, the loss will be larger than a specified limit.

Potential loss is instead interpreted as an unconditional quantity, defined before failure occurs. Therefore, determination of the distribution of the potential losses requires an estimate of the probability of failure, as use of historical data related to the losses from failures can only be used to determine the distribution of the conditional loss.

Potential loss has been defined as a random variable and is therefore characterized by a cumulative distribution function $C(x)$ and probability density function $c(x)$. The probability density function $c(x)$ yields the probability $c(x)dx$ (before failure occurs) that the potential loss X will be in the infinitesimal interval $[x, x + dx]$. This probability is represented as:

$$P(x \leq X \leq x + dx) = c(x)dx \tag{2.1}$$

Similarly, conditional loss (loss given failure) is a random variable and is characterized by a cumulative distribution function $C(x|f)$ and probability density function $c(x|f)$. The conditional probability density function $c(x|f)$ gives the probability $c(x|f)dx$ that the loss X will be in the infinitesimal interval $[x, x + dx]$ given that failure has occurred. This probability is represented as:

$$P(x \leq X \leq x + dx|f) = c(x|f)dx \tag{2.2}$$

The cumulative distribution function $C(x) \equiv P(X \leq x)$ of the potential loss gives the probability that the potential loss X will not be greater than a specified value x. From this, it can be established that the unconditional probability $C(x) \equiv P(X \leq x)$ that the potential loss X will not be greater than a specified value x is equal to the sum of the probabilities of two mutually exclusive events: (1) failure will not occur and the loss will not be greater than x and (2) failure will occur and the loss will not be greater than x [45]. The probability of the first event is $(1 - p_f)H(x)$, where p_f is the probability of failure and $H(x)$ is the conditional probability that the loss will not be greater than x given that no failure has occurred and one assumes the properties of the Heaviside unit step function:

$$H(x) = \begin{cases} 1, & x \geq 0 \\ 0, & x < 0 \end{cases} \tag{2.3}$$

The probability of the second compound event is $p_f C(x|f)$ where $C(x|f)$ is the conditional probability that given a failure, the loss will not be greater than x. The probability that the potential loss X will not be greater than x is therefore given by the distribution mixture:

$$C(x) = P(X \leq x) = (1 - p_f)H(x) + p_f C(x|f) \tag{2.4}$$

By being able to derive the cumulative distribution of the potential loss, a risk model and methodology are presented which allow the formulation of disaster tolerant requirements within the concept development phase of the systems engineering process. Requirements may be set based on an evaluation of the potential losses resulting from critical disaster events.

2.15 Development of Disaster Tolerant Requirements

The need to develop methods for the design of disaster tolerant systems and for augmenting existing large-scale systems with a disaster tolerant capability has been established. While the area of fault tolerance has reached a degree of maturity, disaster tolerant engineering practices are in their infancy. Based on the discussion above, the following design characteristics have been identified for necessary consideration for establishing a methodology and model for setting disaster tolerant requirements based on cost of failure analysis:

1. Determine the critical threshold value (further defined below), based on consequence of failure and threshold of risk investment (further defined below and including cost of intervention, cost of replacement, cost of lost production);
2. Determine the expected value of the potential losses (the risk) from failure;
3. Determine the probability that potential losses from failure exceed the critical value.

Addressing these three design characteristics, a high level process has been defined and is illustrated in Figure 2.13. This methodology will provide a means to incorporate the objectives necessary for integrating disaster events into the systems engineering process.

This requirements development methodology will transform the requirement-driven view of desired disaster tolerant services into a technical specification for systems that provide disaster tolerant capabilities based on expected losses from critical failures.

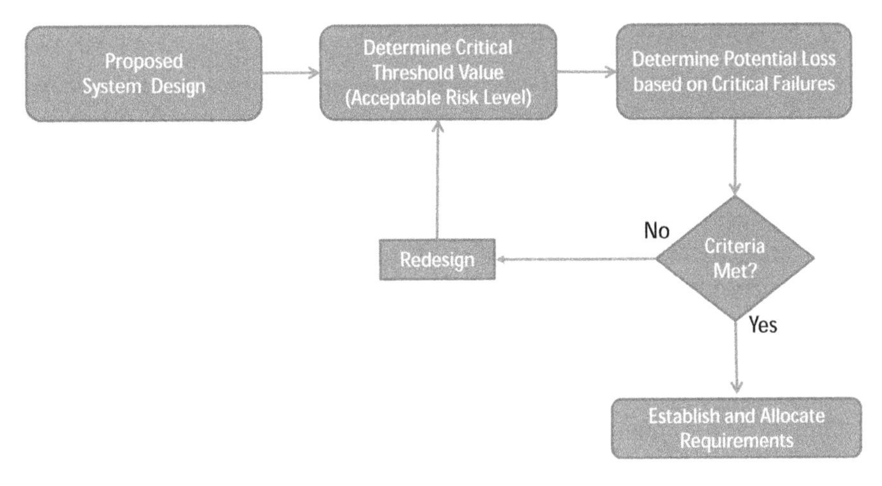

Figure 2.13 Establishing disaster tolerant requirements

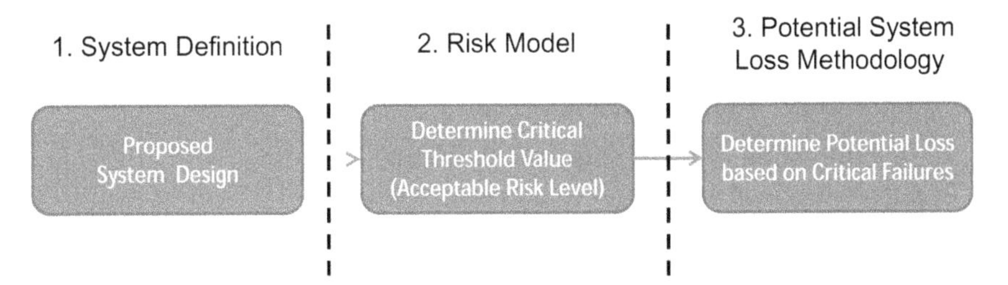

Figure 2.14 Model and methodology overview

2.16 Model and Methodology Overview

Three specific phases have been developed to quantitatively establish disaster tolerant requirements, outlined in Figure 2.14. The first phase is to define the proposed system model. This phase is primarily user driven and provides the basis for system evaluation. The model provides the necessary structure to proceed to phase two and phase three. The second phase is to establish the Critical Threshold Value (CTV). The CTV is the acceptable level of tolerable risk, being a factor of reliability and potential threats the system may be subject to during its life (i.e., if various parts of the system have a physical location in a hostile environment). The third phase is the evaluation of the system based on expected and potential loss from critical failures.

2.17 Phase One: System Definition

The prerequisite for starting the systems engineering process is a user-defined need (or outcome). The purpose of systems engineering in this phase is to define an optimal system concept and concept of operations (CONOPS), describing the characteristics of a proposed system from the viewpoint of an individual who will use that system. The CONOPS is used to communicate the quantitative and qualitative system characteristics to all stakeholders. The Committee on Pre-Milestone *A* and Early-Phase Systems Engineering states that the critical functional elements to consider in the system concept and definition phase include a set of alternatives [46]. The authors of this report recommend at least two alternative concepts be considered before locking in a solution. The second consideration recommended is the "time value of capability." That is, choose concepts that can deliver an initial capability within a reasonable time frame. For example, for major defense acquisition systems, this might be measured as a 3–5 year time frame from Milestone *B* in the Defense Acquisition Lifecycle Framework [46,47].

For any environment the system is intended for use in (defense or commercial) the essential elements for system definition that must be addressed remain the same. These include:

1. Determination of the system stakeholders;
2. Clearly defined system goals and objectives;
3. Understanding of the external constraints on the system;
4. Clear definition of the Concept of Operations;
5. Other clearly defined key system attributes (as necessary).

2.18 Phase Two: Risk Model

Phase two of the developed methodology involves assessing the total risk from failure scenarios. This involves the following steps as illustrated in Figure 2.15:

1. Identify potential failure scenarios;
2. Estimate the probability of occurrence of each failure scenario;
3. Estimate the risk associated with each failure scenario;
4. Estimate of the total risk by accumulating the risks associated with the separate failure scenarios.

The term risk is used in diverse environments resulting in many subtle variations in how it is defined. As a result, there is no universally accepted

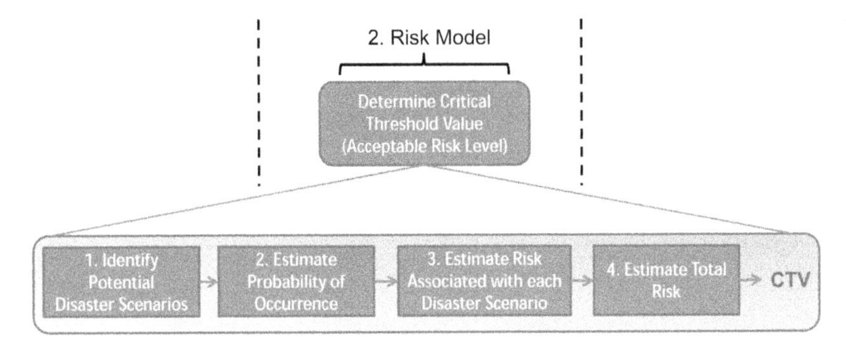

Figure 2.15 Risk model

definition of risk. At the same time, some characteristics of risk are consistent across its many applications. For risk to exist in any circumstance, the following three conditions must be satisfied:

1. There must be a loss associated with a certain situation;
2. There must be some uncertainty with respect to the eventual outcome;
3. Some choice or decision is required.

Most definitions of risk focus on the first two conditions: loss and uncertainty, because they are the two measurable aspects of risk. Therefore, risk can be defined to be the possibility of suffering harm or loss.

Due to its complicated nature, risk can be further divided into two types: speculative risks and hazard risks [48]. Speculative risk allows the realization of a profit, or a positive risk investment. However, this type of risk also allows the potential to experience a loss, making the situation worse off than at a baseline. By contrast, hazard risk only has potential losses associated with it, providing no opportunity to improve upon the current situation. Figure 2.16 illustrates the differences in these two types of risk.

The risk produced by a disaster event is an *extrinsic risk* because it is an underlying trigger (i.e., the occurrence of an event) that occurs outside of expected or predictable operational conditions. The mechanism responsible for generating extrinsic risk (i.e., an event in conjunction with one or more vulnerabilities) also influences its basic properties, which are measured using probability and impact. In general, the probability associated with extrinsic risk is heavily influenced by the likelihood that its triggering event will occur. As has been defined throughout this chapter, disaster events are those with the potential for producing very high, catastrophic consequences and have

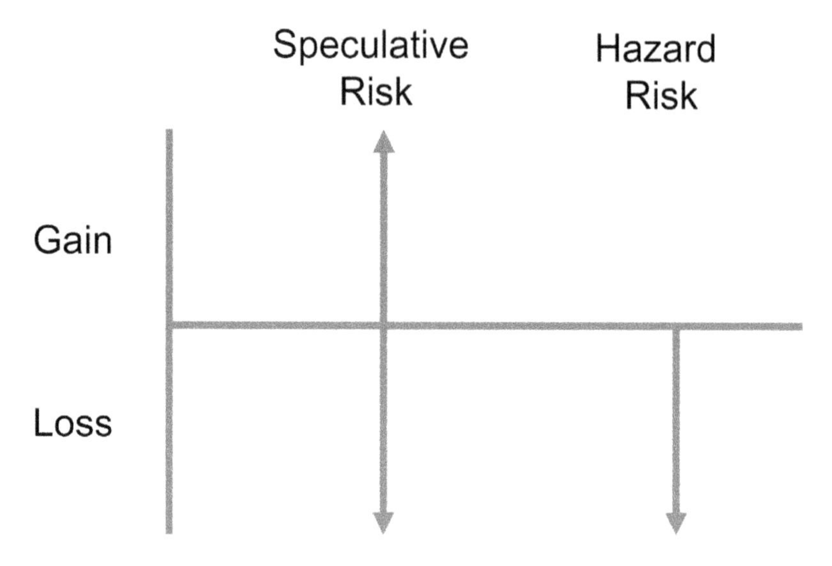

Figure 2.16 Speculative and hazard risk

very low probabilities associated with them. The risk model developed in this chapter focuses on the impact of hazard risks and the potential risk of loss from critical failures.

2.19 Phase Three: Evaluation of Potential System Loss

The ability of a disaster tolerant model to determine if a potentially exploitable element in a proposed system can be used to degrade the performance of an asset, within a system based on multiple system failures, is of great value in today's vastly interconnected society. From this, the potential for risk based on the cost of failure may be reduced by a system designer, and a level of disaster tolerance within the system may be established. The third and final phase in the developed methodology involves the simulation of a proposed system design to determine the expected value of potential losses from critical failures, to determine if the system structure satisfies the level of tolerable risk established in the previous phase. This provides the system designer a quantitative assessment tool and a disaster tolerant based risk mitigation strategy, for system design ensuring operational functionality in the face of disaster events.

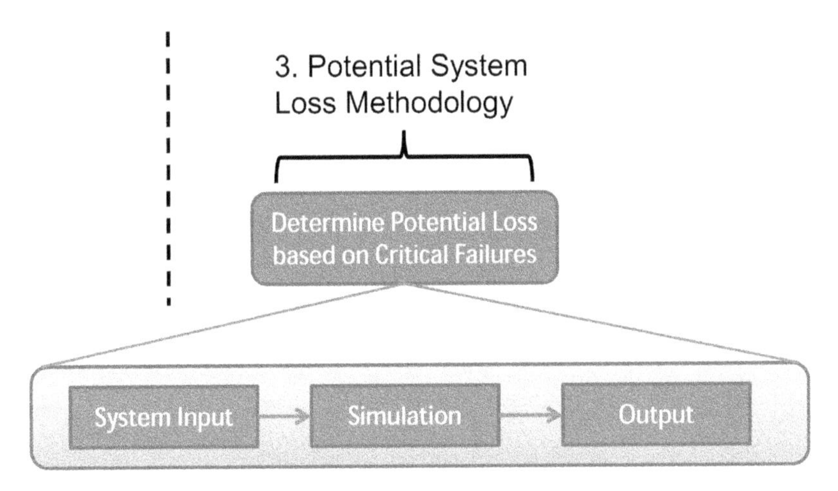

Figure 2.17 Simulation methodology

Simulation is the process of implementing a series of numerical experiments on a system model. An "actual" realization of states is simulated on each system component. During the course of the simulation, component events (working or failing) are made to occur at times determined by random processes obeying failure or repair time distributions of the component. A system (state) realization is then composed according to all component realizations and the system design. After having observed many realizations ("histories") of the system, estimates are made of the desired indices statistically. Figure 2.17 outlines the process for the third phase of the methodology, i.e. the determination of potential loss from failures.

The simulation methodology includes a System Input block that consists of the number of trials to be run and the graph (network structure) to be modeled. The simulation block provides the process of determining the probability of total system failure. The output block provides the determination of whether the network structure, based on the expected value of the potential losses from failures meets the pre-established risk criteria set by the critical threshold value. If $CTV_x > \sum \overline{L}$, the system design meets the disaster tolerant requirements, while if $CTV_x < \sum \overline{L}$, the system must be redesigned.

Therefore, establishing a methodology and a model for setting disaster tolerant requirements based on cost of risk from critical failures due to a potential disaster event requires the determination of three specific calculations, namely:

1. The critical threshold value, based on consequence of failure and threshold of risk investment (cost of intervention, cost of replacement, cost of lost production);
2. The expected value of the potential losses (the risk) from failures;
3. The probability that the potential losses from failure exceed the critical value.

Development of the methodology to achieve these three stated objectives requires an evaluation of available and possible approaches. Examples include Monte Carlo simulation or discrete event simulation. The following section provides the framework for evaluating identified candidate approaches to implement phase three of this methodology.

2.20 Implementation of Phase Three: Determination of Potential Losses

To establish quantitative disaster tolerant requirements, the potential losses from failure are a critical figure of merit. The potential losses can be determined from combining the probability of critical failures occurring in a specified time interval and the magnitude of the losses given that those failures have occurred. This is determined by simulating the behavior of the system during its lifecycle where potential losses are tracked, as illustrated in Figure 2.18. Variation of the number of failures and their time occurrences during a specified time interval (e.g. the design life, which is the period of time during which the item is expected by its designers to work within its specified parameters; in other words, the life expectancy of the item) causes a variation in the potential losses.

The calculated values for the potential losses are subsequently used to build a cumulative distribution. This provides an opportunity to determine the probability that the potential losses will exceed the specified critical threshold set by the system designer, the necessary element for identifying the threshold of risk investment determined by the system designer. Taking the average of the simulated potential losses, from all simulation histories, yields the expected losses from failures. Figure 2.18 illustrates the simulation flow sequence.

A large number of simulations of critical failures histories during the system's life cycle reveals the variation of the potential losses. After finishing all Monte Carlo simulation trials, the cumulative distribution of the potential losses is built. Dividing the sum of all potential losses obtained from the

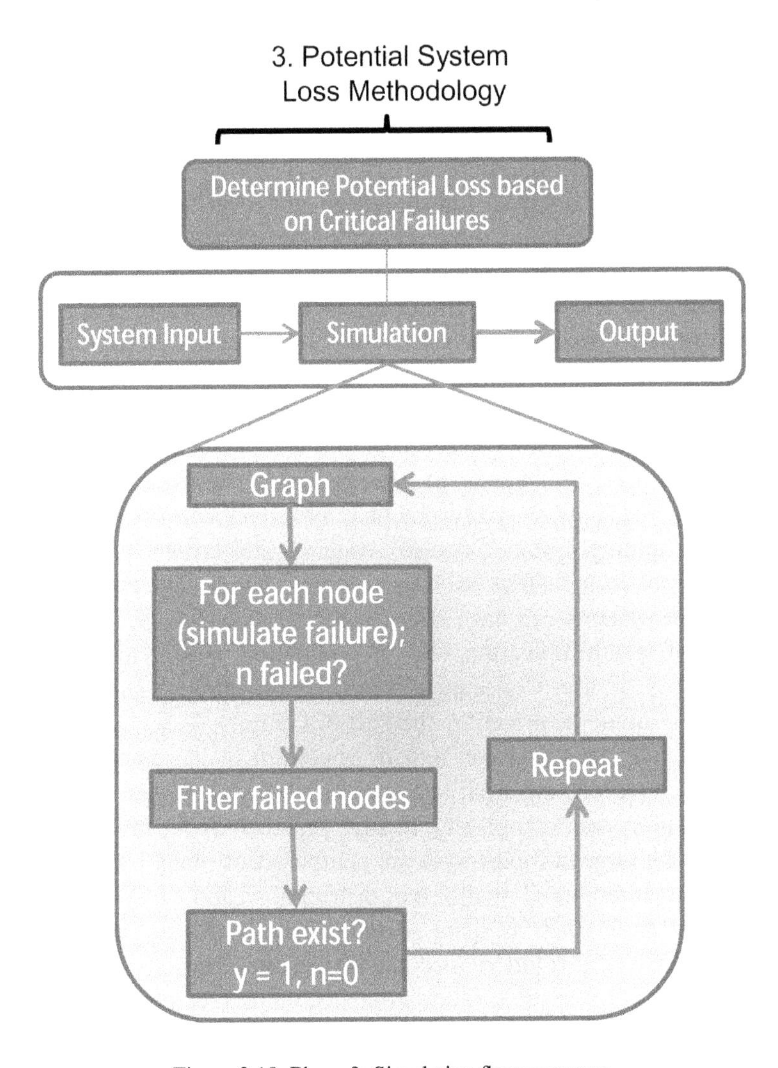

Figure 2.18 Phase 3: Simulation flow sequence

separate trials to the number of simulation trials yields the expected losses from failures. A methodology to determining the effect of multiple system failures in a finite time period, the discrete-event simulation is used, in order to reveal the distribution of the potential losses from failures.

2.21 Conclusions and Direction for Future Work

There is a strategic need to understand the societal consequences of critical failures risks along with benefits of various tiers of increased reliability. From an infrastructure interdependency perspective, power, telecommunications, banking and finance, transportation and distribution, and other infrastructures' are becoming more and more congested and are increasingly vulnerable to failures cascading through and among them. A key concern is the avoidance of widespread failure due to these cascading and interactive effects. Moreover, interdependence is only one of several characteristics that challenge the control and reliable operation of these systems.

With modern technology and higher reliability, availability and survivability requirements, systems are getting more complicated. Complex systems such as multi-level redundant, on-line repairable, networked designs have been seen increasingly in many practical systems, such as communication systems, computing systems, control systems, and critical power systems. Critical failures in industries such as complex defense systems and critical infrastructure systems can have disastrous consequences. Such failures can entail loss of system operation, loss of production, loss of mission success and even, loss of life. Consequently, setting quantitative disaster tolerant requirements should be driven by the cost of failure.

Redundant techniques span a wide spectrum in the design space, and allow a system to achieve high levels of reliability; however, this comes at the expense of system complexity. In such practical applications, where one is dealing with large complex systems composed of many components that fail and get repaired based on different distributions and with additional "real world" constraints (such as spare parts availability, Mean-Time-To-Repair, etc.), exact analytical solutions become intractable. Thus, to solve such systems, one needs to resort to simulation techniques to obtain metrics and results of interest [49, 50].

A methodology and model have been developed which allow the ability to establish a critical risk level prior to system design to determine if the proposed design provides adequate level of disaster tolerance. The output of the methodology and model allows disaster tolerant requirements to be established and allocated to the necessary system elements, to ensure a level of confidence in the event of a potential disaster.

The cumulative distribution of the losses from failures permits estimating the probability that the losses will exceed a critical unacceptable threshold value. The developed methodology executes a discrete event simulation program for tracking the losses from failures based on the Monte Carlo simulation method, to provide the cumulative distribution from which the probability that the losses will exceed a particular value can be readily estimated. This is due to the very nature of the Monte Carlo simulation, which is based on numerous trials, building a statistical distribution describing the various times to failures of any particular loss. Consequently, the Monte Carlo simulation process provides the expected value of the losses from failures.

Disaster tolerance places demands on requirements for architectural design, and system operation and administration. These demands include defining and communicating disaster tolerant policies, monitoring system use, responding to system intrusions and failures, and evolving system functions as needed to ensure disaster tolerance as usage and failure patterns change over time. In this regard, the purpose of proper requirements analysis is to transform the stakeholder, requirement-driven view of the desired services into a technical view of a required product that could deliver those services. This process builds a representation of a future system that will meet stakeholder requirements and that, as far as constraints permit, does not imply any specific implementation. It results in measurable system requirements that specify, from the developer's perspective, what characteristics the system is to possess and with what magnitude in order to satisfy stakeholder requirements.

Disaster tolerant goals are emergent properties that should be designed for the system as a whole, but does not necessarily prevail for individual nodes of the system. This approach contrasts with traditional system designs in which specialized functions or properties are assured for particular nodes and the composition of the system must ensure that those properties and functional capabilities are preserved for the system as a whole. For disaster tolerance, we must achieve system-wide properties that typically do not exist in individual nodes. A disaster tolerant system must ensure that desired survivability properties emerge from the interactions among the components, in the construction of reliable systems from unreliable components.

Dealing with system wide disruptions that may potentially result in infrastructure disasters sometimes requires diagnostic and corrective actions. In almost all cases, minimizing the loss of aggregate value to users and ensuring that it remains within a range required to safeguard the public interest is achieved only by taking a system-wide view, coupled with disaster tolerance techniques and technologies. Modeling for these types of failures in complex systems requires interdependent sector analysis and the ability to handle potentially dissimilar fault models.

It is envisioned that disaster tolerant systems be capable of adapting their behavior, function, and resource allocation in response to a system wide disaster. For example, functions and resources devoted to non-essential services could be reallocated to the delivery of essential services and to intrusion resistance, recognition, and recovery. Requirements for these systems must also specify how the system should adapt and reconfigure itself in response to a disaster or intrusion. Small systems may require few or no essential services and recovery times measured in hours. Conversely, large-scale systems of systems, such as a countries energy infrastructure grid or an international banking system may require a core set of essential services, automated intrusion detection, and recovery times measured in minutes. Embedded command and control systems, such as those typically found in weapon defense systems may require essential services to be maintained in real time and recovery times measured in milliseconds. The attainment and maintenance of survivability consume resources in system development, operation, and evolution. The resources allocated to a system's survivability should be based on the costs and risks to an organization associated with the loss of essential services.

2.22 Summary

In studying large-scale systems with technological, societal, and environmental aspects, the efforts in the modeling as well as in the design and optimization are magnified and often overwhelm the analysis. This is due to the very large number of variables and complexity of the design models. Models must be built to address this complexity, but no single model can ever capture and represent all the essence of large-scale systems.

Traditionally, systems development addresses a variety of design factors. For example, environmental impact, usage, safety, reliability, maintainability, and availability are considered in the system design phase. However, a disaster event has not been included in this manner within the systems engineering design phase. A methodology to include disaster events in the systems engineering process is therefore necessary. For disaster tolerant systems there are obviously high costs of failure. When addressing the necessary elements within the concept development phase, specifically setting requirements that will incorporate disaster events, it is critical to understand that setting requirements based solely on a high availability target does not necessarily limit the risk of premature or early life failure. To ensure proper disaster tolerance in this regard, it is necessary to incorporate provisions for replacing failed components or upgrading performance without disrupting operation, within the design specification to account for premature failures.

Disaster tolerance must be designed into systems to help avoiding the potentially devastating effects of systems compromise and failure due to intrusion or threats against the system. The natural escalation of offensive threats versus defensive countermeasures has demonstrated that no practical systems can be built that are invulnerable to attack. Despite best efforts, there can be no assurance that systems will not be breached, due to either malicious attack or a natural disaster. Thus, the traditional view of systems security and survivability must be expanded to encompass the specification and design of system behavior that help the system to tolerate catastrophic behavior in the face of a disaster. Only then can systems be created that are robust in the presence of attacks and are able to survive these attacks, either natural or manmade, that cannot be completely repelled.

3

Power Supply and Communications Infrastructure Issues in Disaster Areas

Alexis Kwasinski

The University of Texas at Austin, USA;
e-mail: akwasins@mail.utexas.edu

3.1 Communication Systems Infrastructure

Communication networks are a critical infrastructure for nations and society as a whole. They include within themselves infrastructure network elements that play a critical role in communication systems survival during disasters. Thus, communication networks infrastructure elements are interpreted in here as all physical structures necessary for communication systems operation. These communication systems include a wide range of networks, including traditional public switched telephony network (PSTN), wireless communications (also known as mobile telephony networks), cable-TV (CATV) both for TV signal transmission and as a digital telephony carrier, data centers, TV and public radio broadcast stations, amateur radio emergency communications (ham radios), and emergency responders radio systems. All these networks provide essential societal services that become even more important during the rescue, recovery, restoration, and reconstruction phases of a disaster. Typically, the critical nature of communications systems during disasters is associated with emergency operations during rescue and recovery phase, such as E-911 services in the US, as recognized by the International Telecommunications Union (ITU) when it states that "in disaster and emergency situations, telecommunications can save lives" [51]. It is also recognized that good communications are also essential during recovery and restoration operations, by, for example, assisting in coordinating logistical operation of both

N. Marchetti (ed.), Telecommunications in Disaster Areas, 53–94.
© 2010 *River Publishers. All rights reserved.*

emergency responders and critical services operators, such as electric power or water distribution companies. As former United Nations Secretary-General Kofi Annan indicated, "from the mobilization of assistance to the logistics chain, which will carry assistance to the intended beneficiaries, reliable telecommunication links are indispensable" [52]. During the long-term reconstruction phase after a disaster, communication-supported services, such as the Internet, and financial and banking operations, are also very important.

3.1.1 Communication Systems Infrastructure

Although all the varying types of communication networks have important technological differences, in essence their infrastructure is relatively similar. In general, communication networks infrastructures can be divided among inside plant, outside plant, and customer premise's elements. Figures 3.1 and 3.2 can serve to exemplify the basic infrastructure elements found in communication networks. They show simplified schemes of a PSTN central office (CO) and a wireless communications mobile telephony switching office (MTSO) service area, respectively. In both networks, inside plant infrastructure is located at centralized network elements, i.e., the CO or the MTSO. Important centralized infrastructure elements include the CO or MTSO buildings, the power plants, and the air conditioning system. Other centralized network facilities include data centers or CATV head-ends, and TV and radio broadcast transmitters. Communication infrastructure located at the outside plant or at the customer premises are distributed network elements. The main distributed network elements in wireless networks are base stations located in cell sites. Medium to high capacity base stations are typically placed in shelters (e.g., Cell Site #7 in Figure 3.2). In addition to the base station electronic communication equipment, each shelter has a power plant and an air conditioner. Conventional PSTN outside plant network elements do not require local power because they are powered from the CO. However, with the advent of digital communications, more PSTN outside plant network elements, such as digital loop carriers (DLCs) remote terminals (RTs), were linked to the CO with fiber optic cables. Since fiber optics cannot transmit electric power, the conventional power feed from the CO needed to be replaced by local power plants located inside the RTs cabinets (Figure 3.3). Similar power needs are found in CATV optical nodes, line extenders, and bridge amplifiers. Hence, DLC RTs and CATV optical nodes are the equivalent in PSTN and CATV networks of base stations in wireless systems. However, although many base stations require air conditioning, most CATV optical nodes or DLC RTs do

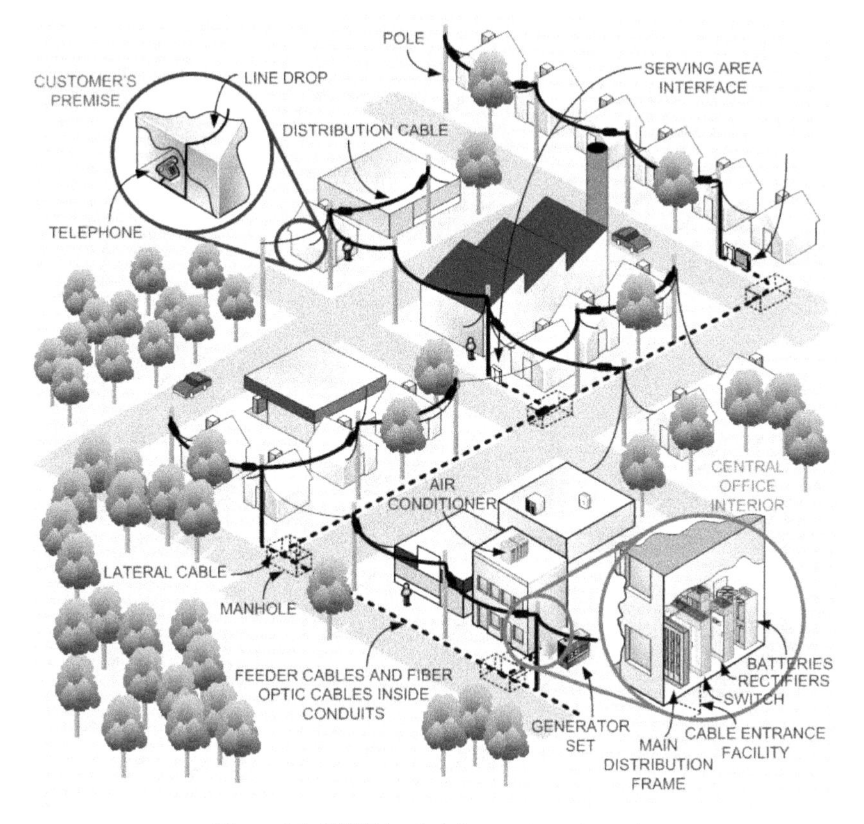

Figure 3.1 PSTN basic infrastructure elements

no need air conditioners, except in the few cases when DLC RTs are placed in huts or vaults.

One important difference with significant implications within the context of disasters between wireless networks, and both PSTN and CATV, is found in the way communication circuits are connected. While in public switched telephony and CATV networks connections between centralized and distributed elements are mostly fixed through copper, fiber optic, or coaxial cables, in wireless networks the connections are more flexible because microwave or satellite links can be used instead of fiber optics and because there is no fix connection between the network and its users. Hence, since wireless networks links can be realized without hardwired connection, and those links can simply be configured through software commands only, mobile telephony

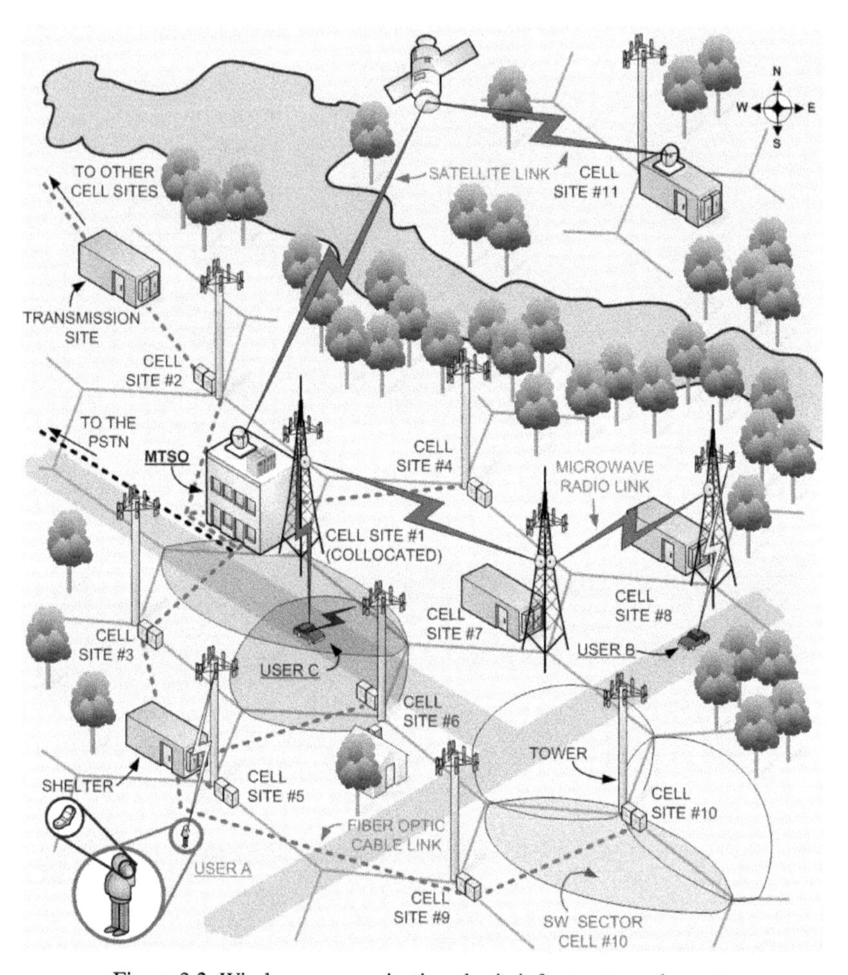

Figure 3.2 Wireless communications basic infrastructure elements

networks have more operational and deployment flexibility than the PSTN. On the contrary, PSTN connections are primarily hardwired, which limits how this network can adapt to critical events. Even when fiber optic in PSTNs provides some more flexibility than conventional copper pair cables – particularly in terms of adapting to varying capacity needs – fiber optic cables still have some degree of inflexibility because they need to be supported by fixed infrastructure elements, such as poles, conduits, termination blocks, and cabinets needing local power. Furthermore, users still have a fixed connection

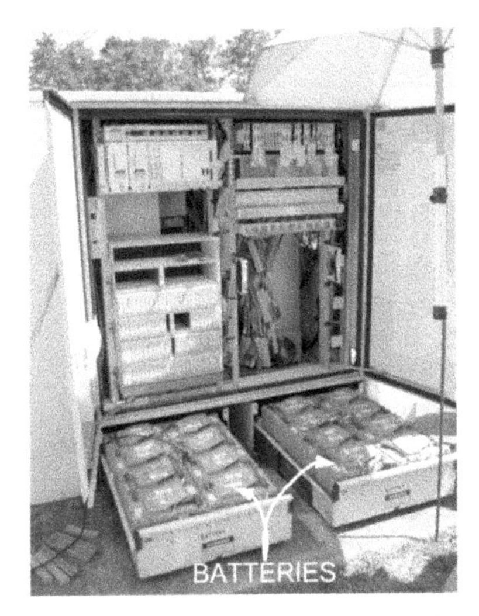

Figure 3.3 (a) Left: Front of a DLC RT cabinet showing the communication equipment on top and two draws for battery strings on the bottom. (b) Right: Side view showing two rectifiers

to the network. In the case of CATV networks carrying digital telephony services, such as voice-over-IP (VoIP) telephony, although there is no hierarchy among communication centers, and its data architecture is almost fully distributed, hardwire connections to the subscribers lead to similar issues from those found in the PSTN.

3.1.2 Communication Systems Power Infrastructure

Highly available power supply is essential for communication systems reliable operation, particularly during disasters. As it is discussed in the next section, power issues cause most communication outages during disasters. If the switch fabric and its controller are the system "brains" and the distributed elements are its "limbs", then the power infrastructure is its "cardiovascular system". A typical communication system node infrastructure elements and electric power architecture is exemplified in Figure 3.4 using a cell site. As shown, electronic communications equipment is fed with dc power (in data centers and many CATV networks, ac is still the predominant power supplied to communications and data equipment) using two circuits in a duplex configuration in order to avoid outages that may happen if a fault occur in an unique circuit. The dc power is obtained by converting the ac grid power

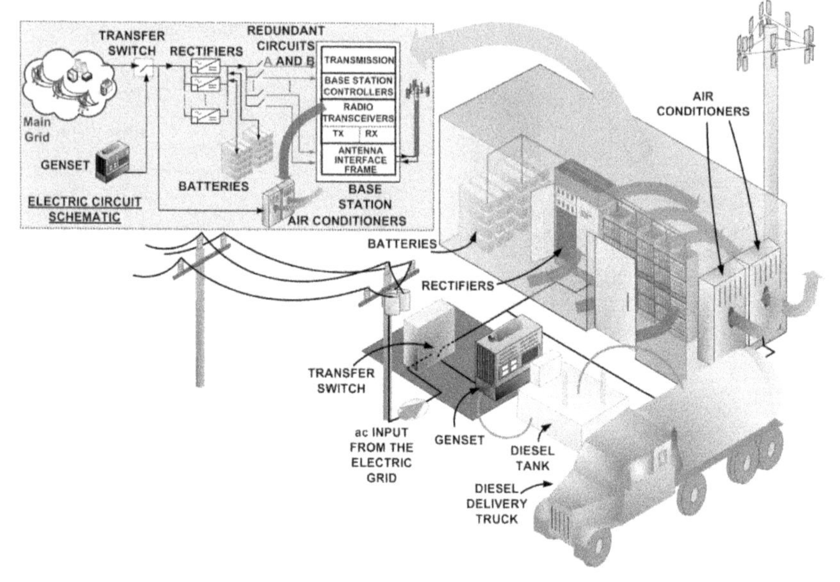

Figure 3.4 Cell site infrastructure layout and simplified circuit schematic in the upper left-side corner. Most communication sites are similar to this example

with rectifiers in an $n + 1$ redundant configuration – in $n + 1$ redundant configurations one additional rectifier is added to the n that are needed at a minimum so in case one rectifier fails there are still n operating rectifiers to power the load. In case of short main outages-a few hours long-the power path composed of the ac input and the rectifiers is backed-up with locally stored energy in batteries. In normal operation, the batteries are kept charged from the electric grid through rectifiers. When the electric grid fails, the batteries start to discharge and keep the communications equipment powered. Depending on the site configuration, batteries can provide power from few minutes up to some hours, until they are discharged. To avoid fully discharging the batteries, most of important communications sites have stand-by diesel generator sets (gensets) which provide power through the rectifiers after a short battery discharge. Less critical telecom network nodes such as low-traffic cell sites and DLC RTs do not have permanent gensets.

Gensets may operate for extended time if they are refueled. Central offices gensets fuel tanks are usually engineered to store enough diesel to keep the facility fully operational for up to 72 hours. However, stand-by generators

Table 3.1 Telecom power plant component availability data in normal conditions. *MUT: Mean up-time, which equals the inverse of λ. **MDT: Mean down-time, which equals the inverse of λ

Item and origin of the value	MUT* (hours)	MDT** (hours)	Availability A
Rectifier [56]	500,000	166.6	0.999667
Genset ($\rho_{GS} = 0.0241$) [56]	823	5	0.9939
Transfer switch [57]	102,041	5.74	0.99994
Power grid tie [56]	2,440	2.08	0.999150

are not intended for long-term operation: a study by the nuclear power plants industry yielded that diesel gensets have an 85% availability when they are operated for more than 24 hours [53]. Although such a low availability may likely be a pessimistic assessment, it can be expected that gensets availability will decrease as the time of continuous operation increases. Gensets long term operation after disasters is also hampered by complicated diesel delivery logistics because in many disasters large areas become inaccessible due to damaged roads. One other issue with gensets is the probability that they will not start when required, usually due to lack of maintenance, for example, by not inspecting regularly the good condition of the genset starting battery.

A better understanding of communication systems power architecture and how communication networks react to disasters can be gained by analyzing the power supply chain availability for a typical site. Communication systems typically require an overall availability of at least 0.99999, or as it is termed: 5-nines [54]. However, reliable electric grids, such as those found in the US have an availability of 3-nines, obtained by replacing the corresponding values from Table 3.1 in

$$A_{PG} = \frac{\mu_{PG}}{\lambda_{PG} + \mu_{PG}} \tag{3.1}$$

where λ_{PG} is the power grid failure rate, and μ_{PG} is the power grid repair rate.

Availability can be improved by combining a diesel genset to the ac feed. The ac power supply availability can then be calculated from [55] as

$$A_{ac} = \left(1 - \frac{(\lambda_{GS} + \rho_{GS}\mu_{PG})\lambda_{PG}}{\mu_{PG}(\mu_{PG} + \mu_{GS})}\right) A_{TS} \tag{3.2}$$

where λ_{GS} is the failure rate of the series combination of the generator set and diesel circuit, μ_{GS} is the genset and fuel repair rate (the inverse of the time that the fuel circuit and fuel provision is down due to failure), ρ_{GS} is the genset failure-to-start probability, and A_{TS} is the transfer switch availability. Considering the typical values in Table 3.1, A_{ac} in normal operating conditions and with a genset with unlimited fuel supply is around 4-nines, which is not sufficient to meet the target availability goal of 5-nines. Without a genset, A_{ac} equals A_{PG} indicated in (3.1), i.e., approximately 3-nines. The main reason for this low availability at the ac tie in both configurations is the power grid low availability that creates a floor value for A_{ac} that is too low to be improved enough, particularly, by using only a genset. Since gensets are normally in stand-by and they usually require some time to start-up and provide power, batteries are added to the system in order to maintain the load powered until the genset starts, or until a mobile genset is deployed to the site where there is no permanent one available. Yet, batteries require a regulated dc voltage, which leads to the need of rectifying the ac input power. The availability of the input power path is, then, the result of the combination of A_{ac} and the availability of the rectifiers A_R, i.e., the product $A_{ac}A_R$.

Typically, the availability of a rectifier is almost 4-nines. In order to improve this value, rectifiers are designed in modules which are connected in $n+1$ parallel redundant configuration. In this way, the rectifiers availability is

$$A_{RS} = (n+1)(1-a_r)a_r^n - na_r^{n+1} \tag{3.3}$$

whereas their failure and repair rates are [58]

$$\lambda_{RS} = \frac{n\lambda_r^2(n+1)}{(n+1)\lambda_r + \mu_r} \text{ and } \mu_{RS} = \frac{2\lambda_r^2\mu_r^n C_{n-1}^{n+1}}{\sum_{i=0}^{n-1} C_i^{n+1}\mu_r^i\lambda_r^{n+1-i}} \tag{3.4}$$

where a_r, λ_r and μ_r are the availability, failure rate, and repair rate of each rectifier module, respectively. Although an $n+1$ redundant arrangement of rectifiers can improve their availability to about 5.5-nines, the availability of the input power path will be limited by the lower value between A_{ac} and A_{RS}, in this case 4-nines from A_{ac}.

Availability improvement when adding batteries can be considered with the augmented Markov diagram (Figure 3.5) with respect to that used to derive (1) in [55]. In this diagram, four states are added to the four considered in [55] in order to represent the reliability behavior of a typical communications system power plant with the rectifier system reliability considered in combination with those from the grid tie-formed by the ac input from the main grid, and the genset. Only three states in Figure 3.5 represent the condition in which the load receives power from the grid or the genset through the rectifiers. These three states form the "working states" set denoted by W. The

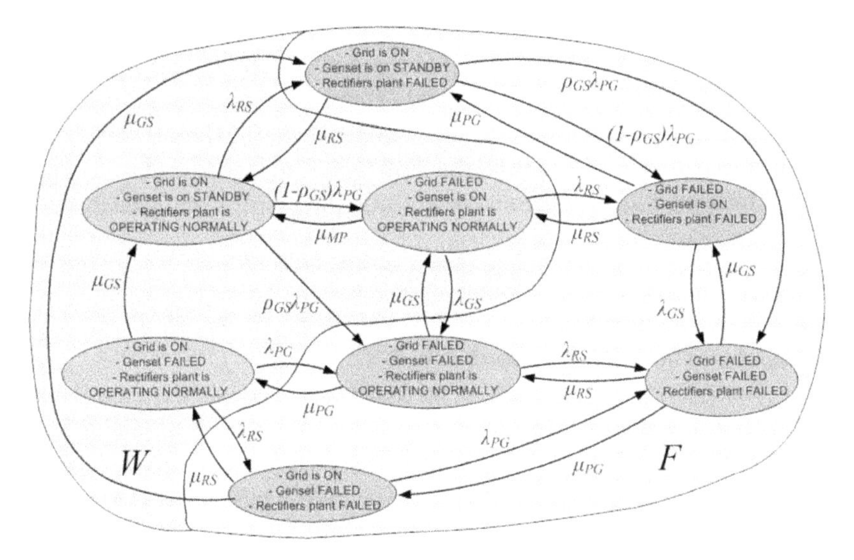

Figure 3.5 Markov diagram representing the operational condition of a telecom power supply path made by the grid-genset pair and the rectification system

remaining five states form the set F that represent the conditions in which the load cannot be powered by the main grid or genset, i.e., the states in which the load cannot be powered through the rectifiers.

From [58] the probability density function $f_{PPf}(t)$ associated with the probability of leaving the set F after being in the set from $t_0 = 0$ and entering W at time t is

$$f_{PPf}(t) = -\mu_F e^{\mu_F t} \tag{3.5}$$

where μ_F is given by all possible transition rates from F to W. That is,

$$\mu_F = -(3\mu_{RS} + \mu_{PG} + \mu_{GS}) \tag{3.6}$$

While operating in the set F the system can only avoid an outage by powering the load from the batteries. If the system operating condition does not evolves into the set W from the set F before the battery backup time T_B passes, the batteries will discharge and an outage will occur. The probability for the system operational condition to stay continuously in F for a time longer than T_B is, then

$$P_{BD}(t > T_B) = 1 - \int_{\tau=0}^{\tau=T_B} f_{PPf}(\tau)d\tau = e^{\mu_F T_B} \tag{3.7}$$

The total power plant system unavailability – the outage probability due to lack of power – is the probability that the power path system portion composed of the rectifiers and the ac power supply components is in the set F *and* that it stays in the set continuously longer than T_B. Hence, based on this conjunction condition denoted by the term "*and*" and the fact that availability is the complement to 1 of the unavailabililty, the power plant availability is

$$A_{PP} = 1 - U_{a,PP} = 1 - U_a e^{\mu_F T_B} \tag{3.8}$$

where U_a is the unavailability related with being in the set F and equals $1 - A_{RS}A_{ac}$. The rectifiers availability A_{RS} is given by (3.3), and A_{ac} equals A_{PG} – from (3.1) – when there is no genset, and equals the expression in (3.2) when there is a genset at the site. In case the batteries discharge and an outage occurs, its expected duration equals the expected value corresponding to $f_{PPf}(t)$ when t exceeds T_B. If for convenience a new variable $x = t - T_B$ is defined, this expected value is

$$T_{OUTAGE} = \int_0^\infty x f_{PPf}(x + T_B) dx \tag{3.9}$$

Thus,

$$T_{OUTAGE} = \frac{e^{\mu_F T_B}}{\mu_F} \tag{3.10}$$

This analysis indicates that with the typical parameters shown in Table 3.1, and with a typical minimum number of necessary rectifiers $n = 7$, at least 3 hours of battery backup is necessary in order for A_{PP} to be over 5-nines; if an outage occurs, power is expected to be restored in 9.5 minutes. With the same number of rectifiers, 6-nines is achieved for $T_B > 6$ hours, and 7 nines is reached with 9 hours of battery backup. In outside plant sites a better value for n is 4. Since in outside plant sites there are usually no permanent gensets, U_a in (3.8) is simply $A_{RS}A_{PG}$ and μ_F is $-(\mu_{RS} + \mu_{PG})$. Thus, at least 8 hours of battery backup, the standard in most DLC RTs in the US, is needed to obtain an unavailability just below 5-nines. However, it is important to emphasize that all these values are observed in normal operating conditions. During disasters, worse reliability performance can be expected.

Previous analysis shows that in normal operation, batteries seem to improve communication systems power supply availability up to the required levels. Yet, the fundamental issue of having the primary supply relying on a single power grid with a relatively low availability remains. In normal operation low grid availability does not lead to significant issues, but during disasters when outages affect extensive areas and portions of the power grid during a long time (as long as several weeks), the grid tie acts as a single-point of failure for communication sites because batteries do not have enough stored energy to power the load for such an extensive time in case there is a failure in the power path through the rectifiers. Moreover, dual feeds do not avoid this single-point of failure because both feeds are typically connected to a same power grid that may fail entirely during disasters. Even more, as Figure 3.4 indicates, air conditioners power supply is not backed up by batteries. Hence, air conditioners availability cannot be higher than that of its power supply. This means that in normal conditions the cooling infrastructure has a power supply availability of 4 nines when a genset is present or 3 nines when there is no genset at the site. Hence, communication sites availability is still limited by the power grid's availability, now indirectly through the cooling infrastructure. If the power grid and the genset (if it is present) fail, thermal inertia – the delay observed in all bodies to increase or decrease its temperature when heat is provided or removed, respectively – may provide to the cooling circuit the same function than batteries do for the electric circuit: to delay communication equipment overheating failure for some minutes and even a few hours. In addition to power grids, communication sites rely on other infrastructures – sometimes called lifelines – in order to maintain continuous operation during long outages. Hence, during disasters there exist interdependencies among critical infrastructures that may make communication systems more susceptible to failure. In addition to the power grid, transportation is one of such critical interdependent infrastructures necessary to support diesel fuel supply logistics for the gensets, as exemplified in Figure 3.6. In some cases water supply is needed for air conditioner chillers. In some other cases, e.g. when using natural gas gensets, natural gas infrastructure may play an important role in maintaining the communication equipment powered. Thus, analysis of communication infrastructure behavior during disasters is a very complex challenge that can be somewhat clarified by discussing the effects of recent disasters on communication systems.

3.2 Effects of some Recent Disasters on Communication Networks

Disasters can be characterized based on their origin: natural and man-made. The latter can be accidental or intentional. Some natural disasters such as tropical cyclones (also called hurricanes in the United States or typhoons in

Figure 3.6 A damaged access road onto a cell site collocated with a silo after the 2010 earthquake in Chile

eastern Asia), floods, winter storms, large thunderstorms and tornado outbreaks, solar storms, volcanoes, and wildfires can usually be anticipated a few days earlier, whereas earthquakes and tsunamis do not provide any appreciable warning time. These disasters may have direct effects, i.e., damage to communication equipment, such as a switch fabric, servers, or base station communications components, or indirect effects, e.g., communications traffic congestion, or outages when no direct damage is caused to the communication equipment, such as communication services loss due to lack of power originated in generator engine fuel starvation, or damage only to communication systems ancillary infrastructure, including the power plant, but not to the main communication equipment.

Direct effects are, usually, more preventable and easier to avoid than indirect ones. For example, as Figure 3.7a exemplifies, in order to avoid damage due to a storm surge – a water inflow from the sea carried inland by a tropical cyclone – DLC RTs have been installed on platforms in areas that were affected by Hurricane Katrina. As both Figures 3.7a and 3.7b show, the same strategy is used in the US in cell sites located in coastal areas prone to hurricanes. In areas prone to earthquakes, reinforced constructions, such as the tower in Figure 3.8a, allows avoiding serious damage to communication infrastructure. Damage to rectifiers due to lightning strikes from thunderstorms,

Figure 3.7 (a) Left: DLC and base station on platforms to avoid damage from hurricanes storm surge. A small boat carried to the site by hurricanes Gustav or Ike is observed behind the tower. (b) Right: Another base station on a platform, with concrete reinforced pylons. The debris line marks Hurricane Ike storm surge height

or high geomagnetic currents induced during solar storms can be prevented with high energy surge arresters instead of simply using standard metal-oxide varistors (MOVs), such as the one in Figure 3.8b.

Disasters indirect actions tend to lead to more serious effects than those from direct actions. Typically, disasters indirect effects are more critical when they affect infrastructure that are needed by communication systems to operate, such as the power grid, and diesel fuel storage, procurement, transportation and delivery infrastructure, in a more severe way. Disasters indirect effects are also more important when they disrupt communication network operators' logistical efforts to repair the damage and to keep all undamaged sites operating. This increased strain on logistical operations worsens as the disaster effects last longer. Clearly, as the number of affected sites increases, the logistical needs in order to avoid outages in these sites also increase. Hence, disasters affecting many infrastructures necessary for the correct operation of communication networks for a long time and affecting an extensive area typically deserve more attention that disasters that do not have those characteristics. Certainly there are always exceptions to this observation, like it happens when a single important facility, such as Verizon's central

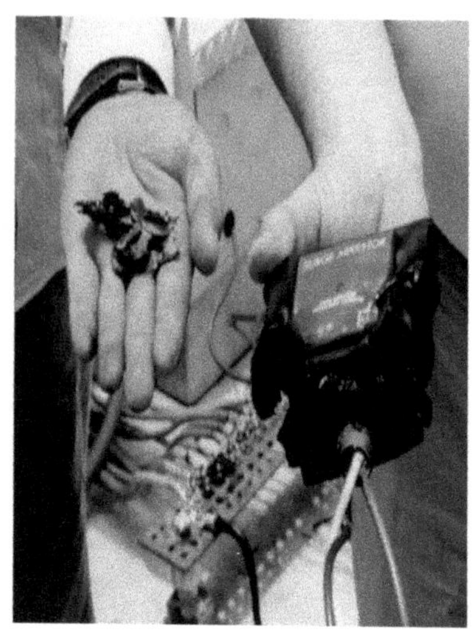

Figure 3.8 (a) Left: An earthquake reinforced communications tower in Tokyo, Japan. (b) Right: Destroyed MOV due to a high-energy surge

office building located next to New York City World Trade Center's (WTC) Twin towers former location, is affected. Another exception is potent volcano eruptions in islands. Thus, in general, the disasters that tend to have a more severe effect on communication systems are tropical cyclones, earthquakes and tsunamis, and in some cases floods. This is the reason why the focus of the following discussion tends to focus more on these critical events.

Since the beginning of the 21st century communication outages originated in disasters have attracted increased attention. In part this increased attention is originated in the more extensive use of digital technologies that, from the user perspective, led to more prevalent communication environments, and, from the service provider perspective, to more distributed architectures materialized through DLCs in the PSTN and base stations in wireless networks. Some earlier reports of the effect of intense storms on communication networks date back to the late 1980s when the U.K. experienced high-wind storms in 1987 and 1990, and in 1989 when Hurricane Hugo struck US Eastern Coast and Northern California was struck by the Loma Prieta earthquake. In all these cases power issues were the main cause of communication outages including 6 failed COs out of the 154 that lost power in the Loma Prieta earthquake [59], one CO lost during Hugo [60], and 6 COs lost during UK's 1987 storm [61]. In addition to these outages in large telecommunications nodes, 57 small telecom centers (less than 600 subscribers) failed during this last storm and 14 similar nodes lost service during UK's 1990 storm because of battery exhaustion. Similar outcome was likely observed in 555 DLCs that lost power during Hurricane Hugo, whereas only 10 DLCs were destroyed [60]. During the Loma Prieta earthquake it was observed that batteries at many DLC RTs discharged faster than expected due to their shorter life caused by high temperatures inside the RT enclosure during normal operation [62].

The next recorded notable disaster was Hurricane Andrew in 1992. No CO lost service due to power related causes, but more than a 1000 DLC RTs lost power, 722 of them in Miami [63]. Due to Andrew's very intense winds, damage to the outside overhead plant was important. Yet, it was not extended over a large area since Hurricane Andrew was not a very large storm. Although the next significant hurricane to affect the US was Isabel in 2003, several earthquakes and other disasters occurred in between. In 1994 the Northridge earthquake impacted primarily the City of Los Angeles, California [64]. Although 2 to 9 COs of the 80 in the area received moderate to severe damage, none lost service. Damage to outside plant cables was also minor. Thirty-five cell sites were affected, most of them due to transmission links issues, e.g., microwave antenna misalignment, and of them only 3 due to power issues. However, switches failed at 4 PSTN COs when the power went out and genset problems happened. These failures affected 224,500 of the 6.5 million PSTN lines in the area, with outages lasting between 1 and 13.5 hours. Air conditioners in some sites were affected by lack of public water supply. Another important earthquake happened almost a year later, this time affecting the City of Kobe in Japan. In total, 285,000 out of 1,440,000 PSTN lines lost service due to power issues [65]. Damage to gensets fueling circuits and anchors led to outages in 2 COs. Another CO lost service because the genset failed to start due to lack of water. All these sites were restored within 36 hours with portable generators. Also, 3 large transmission towers were damaged at other COs. Additional 193,000 outages were caused by damage to the outside plant, both overhead and buried infrastructure [65]. Most of these outages were restored within 2 weeks.

Three noticeable earthquakes happened in 1999, two of them, one in August and the other one in November, affected Turkey. In the first one 10 of the 25 PSTN COs in the most severely affected area lost service due to lack of power when batteries were toppled or gensets were damaged from ground shaking. In many cases power issues also affected equipment cooling. Service in these sites was restored within 3 days. In addition, 1 CO lost service due to building damage and 4 small remote switches failed due to building collapse [66]. Wireless communication networks experienced similar failure rates and causes than the PSTN. There does not seem to be quantifying statistics about the second earthquake in Turkey in 1999, however communication outages were also widespread and at least 1 CO was inoperable due to building damage [66]. Another earthquake affected Taiwan in September [67]. Although 51 COs were affected by the earthquake and its aftershocks leading to 190,000 lines out-of-service (OOS), only one of those (Nantou) was significantly damaged – its operations were moved to the undamaged building next door. In the rest, mostly with small remote switches, power issues were the leading cause of failure. Most of these COs were restored within 2 days with mobile generators although a few of them took up to 25 days to have their service restored. Almost all the wireless communication networks lost service when 1500 cell sites failed due to lack of power because there were no permanent generators at the sites. A few cell sites were lost due to damage, too. Nationwide, communications were disrupted when some transmission links were interrupted due to tower damage or fiber optic cables severance. These links took up to 3 weeks to repair.

Many powerful earthquakes occurred during the first decade of the 21st century, but in not all of them the impact on communication systems have been studied or documented. In some of these cases, such as the earthquakes that affected Iran, the lack of information may have been caused by security limitations. In other cases, such as the 2002 Alaska earthquake [68] or the two 2003 El Salvador earthquakes [69], infrastructure was well constructed which resulted in minimal damage and a reduced number of outages. The most significant and damaging earthquake was the Sumatra-Andaman Islands earthquake and consequent Indian Ocean Tsunami of December 26, 2004. Although there is no complete information about the effects on communication systems throughout such an extensive affected area it has been reported [70] that in Sri Lanka 25 COs and an unspecified number of towers were destroyed. In the south India peninsula at least 3 COs flooded. This flood resulted in damage to the power plant that had been located at a lower floor because of its batteries heavy weight. Damage to coastal wireless networks and PSTN infrastructure was also important. In all Thailand's area affected by the Tsunami, damage was mostly limited to outside plant infrastructure along the coast because all PSTN COs and most wireless cells sites were located inland. However, in all affected areas by the tsunami, power issues still contributed to many communication outages [59]. Other important recent earthquakes include the 2007 Chuetsu earthquake in which although 6 COs lost power from the grid and they were not equipped with permanent genset, they did not lost service because mobile generators were deployed on time. A similar strategy was used in the 2008 Iwate-Miyagi inland earthquake when 11 of NTT's COs lost power from the grid. In this event, 1 CO (Kouei) went OOS because damaged roads prevented the deployment of a mobile generator by land. During the 2006 earthquake in Hawaii, power problems were the main reason why some wireless communication networks lost service in close to 50% of their cell sites [71]. In this earthquake, damage was little and no outage was observed in the PSTN. In the 2007 Peru earthquake communications were affected by a severed fiber optic cable. Although there is no information about outages in PSTN COs, there was damage to the outside plant and 60 base stations from one of the operators lost service, in many cases due to lack of power when gensets could not be refueled due to damaged roads [72]. A very strong earthquake affected China in 2008. Reports [73] indicate that 616 PSTN COs and 16,500 base stations were damaged or lost service due to lack of power. Outside plant and transmission infrastructures, including 11,000 km of fiber optic cables, were severely damaged. In some cities PSTN operation was restored within 2 to 3 days, but in some large rural areas wireless communications service took up to 60 days to be restored. Still, even if the network had been repaired sooner, users would not have been able to make calls because power outages lasting 4 to 6 weeks prevented charging their hand sets [74].

The most recent relevant earthquakes struck Indonesia in September 2009, Haiti in January 2010, and Chile in February 2010 (the fifth most intense earthquake ever recorded). Although reports are very preliminary, at the time of writing these lines many communication outages occurred in all these 3 events. In Indonesia, outages lasting at least 6 hours have been reported in base stations and remote switches due to damage and power outages. In Haiti, although many cell sites had permanent generators to avoid outages caused by the traditional unreliable power grid in the country, the entire network of wireless operator Voil lost service when a fallen wall damaged the fuel circuit of the main MTSO and render the genset inoperative. Although the site counted with batteries for several hours of operation, the switch at the MTSO was turned off to avoid damage to its components due to over-temperatures. Once the service to the MTSO was restored, damaged roads and the main port complicated fuel resupply logistics at most of Voil cell sites. In Chile, power outages were an important cause of communication outages which in some cases lasted several days. The reason for power-originated outages was that network operators only had permanent generators in a very small portion of their cell sites or DLC RTs. Moreover, even many small switches lack permanent gensets. For example, one network operator lost service due to power issues in almost 1000 cell sites and 200 DLC RTs, Due to damaged roads and difficulties in procuring portable gensets, service in some cell sites was restored more than a week after the earthquake, when electricity from the power grid was once again available. In addition, 150,000 wireline subscribers were affected mostly in small remote switches (fewer than 5000 subscribers) due to lack of power because these sites did not have permanent generators, either. At least 3 larger COs lost service due to collapse walls or damage caused by tsunami waters. A few others sustained damaged, but this damage was not significant enough to prevent operation. At least another CO lost service due to over-temperature when the air conditioner stop working because its genset failed. Although many outages originated in collapsed buildings pulling down drops running between poles, the use of concrete poles prevented even more damage and outages in the outside plant. Many cell sites – most of them located indoor, in shelters – of a network operator also sustained damaged due to improper anchoring of batteries or due to damaged or bended equipment frames. Additional loss of service was cause by fallen or misaligned antennas in about 50% of the sites belonging to that same network operator. Damage to towers was minor: only 2 of them felt, although a few cell sites on building rooftops needed to be relocated because of the extensive damage suffered by the building where they were installed.

The US experimented several significant hurricanes in a 5 year span from 2003 to 2008 with critical impact on communication systems. The first of those was Isabel in 2003. Isabel affected the mid Atlantic eastern coast of US causing significant power outages that affected communication systems. Although no CO experienced outages, more than 800 DLC RTs lost power – most of them likely failed – and 20% of cell sites in 2 wireless networks failed due to lack of power [75]. Four important hurricanes affected the US in 2004, Charley, Frances, Ivan, and Jeanne. Although there does not seems to exist a complete analysis of the effect of these storms on communication systems, each caused around 1 million PSTN outages and 30% of wireless networks loss of coverage; these last ones mostly due to power issues as a very small percentage of cell sites were damaged. Although damage to outside plant and transmission infrastructure led to failures, power issues were also an important cause of outages. In particular, power issues caused failures of many DLC RTs, including near 1,100 of them during Hurricane Frances and at least 397 of the 1,193 DLC RTs affected by power outages during Ivan. Power issues during Ivan were aggravated because most portable generators were still in use in other areas affected by earlier hurricanes in the season. Thus, it is likely that almost all 1,193 DLC RTs that lost commercial power during Ivan were unable to sustain the loads only with batteries. In general, for the four storms wireless service was restored within few days and most PSTN lines regain service within a week.

In addition to Katrina (which will be discussed in separate paragraphs due to its importance), 3 hurricanes affected the US in 2005 – Dennis, Rita, and Wilma – and 3 in 2008 – Dolly, Gustav, and Ike. Hurricane Dennis did not affect communications significantly. Still, 82 of 300 DLC RTs lost service before portable gensets could be deployed to avoid fully discharging the batteries [63]. Wireless service kept at least 80% of the coverage during the storm [76]. Rita had more severe effects to communication systems. Four PSTN COs (one of those shown in Figure 3.9) and 24 DLC RTs were destroyed, and outside plant infrastructure was severely damaged, especially in the coastal areas. Many outages were caused by lack of power in close to 700 DLC RTs [77] [78]. Wireless networks were not so severely affected as the PSTN, with coverage maintained above 80% thanks to the widespread use of permanent genset at the cell sites. Lack of power was also the primary cause for the at least 1 million PSTN lines OOS after Hurricane Wilma, originated in 1714 DLC RTs and 1 small CO that failed without receiving damage, as it did occur with up to 8 DLC RTs [63]. Power failures also originated outages in wireless networks with coverage dropping to less than 60% in some densely populated areas.

The first hurricane that affected the US in 2008 was Dolly. Its effects were minimal with some few DLCs RTs and cell sites loosing service due to power issues. Even fewer distributed network nodes were damaged during the hurricane. The next hurricane to make landfall on the US coast that year was Gustav, which affected the same area struck by Katrina 3 years earlier. Thanks to many measures implemented after Katrina in order to harden the infrastructure, such as placing many DLC RTs on platforms and equipping them with permanent natural gas generators (Figure 3.10a), reports of outages were relatively minor. Still, damage assessments discovered at least one CO with likely issues in its permanent genset as the presence of an emergency mobile generator shown in Figure 3.10b seems to indicate. The last hurricane to make landfall on the US was Ike. With Ike, 1 PSTN CO, shown in Figure 3.11, lost service due to damage [79]. Two COs and likely 4 more located in the most severely affected coastal areas failed due to power issues, although these 4 last ones may also have lost service due to a severed transmission link. Many more outages were caused by failure of 788 DLC RTs when portable gensets

Figure 3.9 Sabine Pass CO. The old building was flooded by Rita and later destroyed by Ike. When the CO was flooded, service was initially restored with the DLC on the right, also destroyed by Ike. A remote switch, on the left, was later added to the site

Figure 3.10 (a) Left: DLC RT with a natural gas genset on a platform. (b) Right: A large mobile genset powering a CO after Hurricane Gustav

Figure 3.11 AT&T Sherwood CO after Hurricane Ike. (a) Left: Front. (b) Right: Back with switches on wheels used to restore service

Figure 3.12 (a) Left: Damaged SAI. (b) Right: Damaged DLC RT

could not be deployed at the sites before the batteries were discharged [79]. Also, 70% of a CATV operator's network with digital telephony services was OOS from lack of power. Outside plant also received damage, particularly in the coast. As Figure 3.12a exemplifies, service area interfaces (SAIs) – also known as feeder distribution interface, a passive outside plant cabinet where feeders with up to a few thousand copper pairs are connected to distribution cables with no more than a few hundred copper pairs – were vulnerable and many were destroyed. Yet, less than 3% of DLC RTs were destroyed, as it happened with the one shown in Figure 3.12b. Although wireless networks experienced complete outages in coastal areas due to damaged based stations (Figure 3.13a) or power issues (Figure 3.13b), in general they showed an adequate coverage level. No tower damages were identified during damage assessments. It is important to point out that once Ike moved inland its low pressure center got combined with a front and created a powerful storm hun-

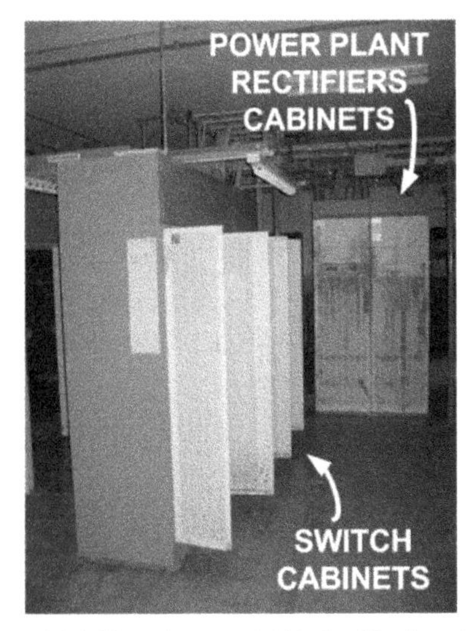

Figure 3.14 Two destroyed COs by Katrina. (a) Left: Mud marks are visible on the cabinets. (b) Right: Yscloskey CO

lost grid power, when locally stored diesel exhaustion led to genset engine fuel starvation, and the engine could not be refueled due to disrupted diesel fuel supply or obstructed roads. Although this last set of COs were restored within a week, the remaining COs outages lasted up to several weeks. A positive note was keeping New Orleans Main CO and tandem switch – a switch that commutates calls among other switches and not among subscribers – operational thanks to armed – guarded delivery of diesel for the generator and water for the air conditioners.

PSTN failure led to many wireless network outages when calls could not be routed. Only a small portion of the 3000 cell sites in the affected area lost service due to damage, most of them because they were not placed above the flood plane. Lack of uniform construction practices with some base stations located above the flood plane while others at the same site located on the ground contributed to this failure mode. Most other outages in wireless communication systems were caused by lack of power when permanent gensets could not be refueled or portable gensets could not be deployed. Since MTSOs are not constrained in their distance to the users, most were located inland and avoided damage. Still, one MTSO failed when it flooded. Although some areas experienced complete loss of wireless communication signals, most of the coverage was restored to pre-storm levels within a week.

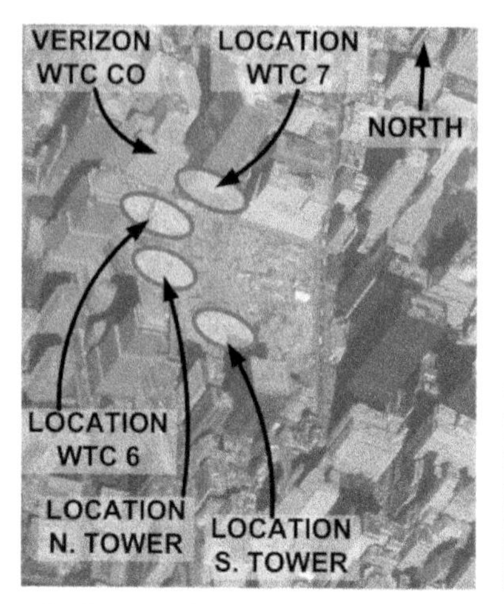

Figure 3.15 (a) Left: Photo showing Verizon's CO at the WTC surroundings. (b) Right: Verizon CO exterior still showing some damage from the attacks 4 months later

A flooded main transmission node in New Orleans also affected communications, and lack of power affected others. Digital telephony through CATV networks was also severely impacted from flooded network facilities, failed network elements power supply, destroyed outside plant optical nodes, and up to 15% of aerial cables damaged [82]. Other essential communication means during disasters such as public radio and TV experienced widespread loss of service due to grid power outages as well as direct damage from wind and storm surge.

Other events different from hurricanes and earthquakes may also have a relatively important impact on communications. One of those was the attacks on New York City's WTC. During the attacks two transport nodes at the WTC buildings were destroyed. In addition two switches failed when their CO building, shown in Figure 3.15, was damaged from falling debris from the WTC North Tower and WTC #7 building collapse [83]. Additional outages were caused in 2 switches by lack of power and by damage to buried outside plant infrastructure. This last failure mode seems to be relatively common during attacks, as exemplified by the 1994 bombing of a Jewish center in Buenos Aires that also destroyed a main route of buried conduits under the explosion. Outside plant infrastructure damage is a leading cause of outages in other disasters like wildfires and winter storms. Yet, power outages are still a failure cause as exemplified by the 2003 California wildfires [84].

In summary, all these events demonstrate that lack of power is a leading cause of communication outages during disasters. Outside plant infrastructure damage is another important contributing cause for outages. Damage to communication towers or buildings tends to me more localized and scattered in nature. Thus, except for some disasters, the impact of direct damage to facilities tends to be, in general, less significant than power-related outages. These issues also manifest the inherent vulnerabilities in the air conditioning systems. The description of these events shows not only that there is a potential single point of failure for communication systems at the grid tie but also that telecommunication systems are not self sufficient. With the technology used nowadays, communication facilities depend on well organized logistical operations, and on other infrastructures, their lifelines. Hence, planning contingency plans in case of disasters is as essential as implementing adequate construction and engineering practices in order to reduce the direct impact of a disaster on communication systems infrastructure. The effects of these disasters show that there does not seem to be a significant difference in whether the disaster can be forecasted a few days in advance or not.

3.3 Restoration Strategies

3.3.1 Public Switch Telephony Network

Digital loop carrier systems are usually extensively used during PSTN restoration operations after disasters. Particularly in the US, DLCs had been used to restore service in destroyed CO areas after hurricanes Katrina and Rita, as Figures 3.9 and 3.10a exemplify. With this approach, surviving switches take over the surviving demand of a destroyed switch by hosting a few RTs located in the service area of the destroyed CO. These DLC RTs are connected to the new host with a fiber optic cable. Use of DLCs to replace damaged COs has some advantages, particularly in areas where a significant portion of the demand has also disappeared as a result of the disaster. One of these advantages is DLC's relatively low cost. Another is the flexibility that DLCs provide to adjust network to an uncertain demand because the fiber optic cable capacity depends mostly on the electronic equipment at the terminals. Hence, with DLCs the cost of installing expensive feeder copper cables with a fixed capacity is avoided. However, use of DLCs has some disadvantages, such as lower network availability, originated in part by power issues that worsen as the number of DLCs increases because of the increased burden on logistical operations. As it was discussed, in some cases DLCs battery backup time is not enough to prevent loss of service during long power grid outages so RTs require the deployment of portable gensets or some other power alternative for long grid outages. Another availability problem is that DLCs are designed with lower target availability than the switch fabric. Yet another disadvantage is that few COs may become a single point of failure for extensive areas because more traffic is then routed through few host COs. Finally, when compared with switches, DLCs can only implement basic functionalities. All these disadvantages may become an important issue in future disasters, particularly when the original plan of installing DLC provisionally results in a permanent installation as exemplified through St. Bernard CO flooded by Katrina and shown in Figure 3.10a. Therefore, a better solution to replace damaged switches is to use "switches-on-wheels" (SOWs). These are mobile switches mounted in shelters or some other protected environment that are transported to the affected site by trucks once the roads allow for such a load to pass through. This is the solution implemented to restore service to Sherwood CO after Hurricane Ike destroyed it. However, in some intense disasters, such as earthquakes, it may take several days to have the roads cleared to allow such trucks to pass. Two important advantages of SOWs over DLCs is that switches provide better functionality when connecting trunks, a requirement for cellular networks links. SOWs also reduce potential congestion points by alleviating traffic of switches that otherwise will be hosting DLC systems. In addition to direct initial capital costs, SOWs have a considerable maintenance and financial depreciation costs, because the system needs to be updated regularly with the latest software versions and the batteries need to be kept constantly charged while the equipment remains idle waiting for a disaster to happen. Also in the US, DLCs are used regularly in order to replace optic cables feeders that were damaged by water immersion in hurricanes or broken conduits during earthquakes. In some cases DLCs are mounted on trailers (Figure 3.16a) or on pallets (Figure 3.16b) for fast deployment. This solution has the same advantages and disadvantages than when DLCs are used to replace damaged COs. Yet, extended use of DLCs in this role may significantly increase logistical needs in future disaster and maintenance costs in normal operating conditions throughout the year.

Service restoration to sites affected by power issues is almost always done with portable generators. In case of COs these portable generators, such as the one in Figure 3.10b, might be bulky, so delays in their deployment may exist if access roads to the affected site are flooded or blocked, as it happened with NTT's Kouei CO after the Iwate-Miyagi inland earthquake. In the US power supply restoration to DLC RTs involve deploying a portable genset,

Figure 3.16 (a) Left: DLC RT on a pallet. (b) Right: DLC RT on a trailer

Figure 3.17 A DLC RT powered by a mobile 10 kW genset

such as the one in Figure 3.17, before the batteries discharge. Although RTs batteries are usually sized for 8 hours of back up, high temperatures typically found in outside plant enclosures reduce battery life and shortens the back up time. In addition to potential complications for portable genset delivery due to blocked roads, their use create a significant logistical undertake because gensets need to be refueled periodically. Hence, in some cases, such as hurricanes when it may not be safe to deploy portable gensets for several hours, outages due to lack of power may not be avoided. One alternative implemented in the area affected by Hurricane Katrina is to locate a permanent natural gas genset in some critical DLC RT sites. These sites performed well during Hurricane Gustav, however cost limitations prevents to extend this solution to all sites. Moreover, this alternative is not suitable for earthquake-prone areas because natural gas service is often interrupted during earthquakes in order to avoid fires.

3.3.2 Wireless Communication Networks

Since wireless communication networks are not limited as the PSTN by hardwired connections, service restoration can be done faster by relying on mobile nodes. Coverage loss is usually restored temporarily with cell-on-wheels (COWs) or cell-on-light-trucks (COLTs) until a permanent fix can be performed by replacing damaged electronic components. As Figure 3.18 indicates COWs and COLTs are base stations mounted on a platform, or inside a trailer, or a truck. In order to avoid power issues, COWs and COLTs are often equipped with their own gensets or are deployed with portable generators. In some cases (Figure 3.18a) COWs and COLTs are collocated at a damaged site and utilize the existing infrastructure such as antennae tower and fiber optic links, but in some other cases (Fig 3.18 (b)) they use their own post and establish satellite links; although some wireless systems may see longer communication delays than others when using satellite links. In terms of costs, COWs and COLTs do not have such a relative high year-round maintenance and depreciation financial cost like the SOWs because they tend to be used more often even when there are no disasters in order to increase coverage and network capacity in specific areas during given periods of time, such as sport events. Regarding loss of service in cell sites without a permanent genset and that experience an extensive grid outage, the most common solution is to deploy portable gensets to the site. Like it happens with PSTN's DLCs, deployment of portable gensets may be complicated due to damaged roads. Even more, since base stations have higher power consumption than DLCs,

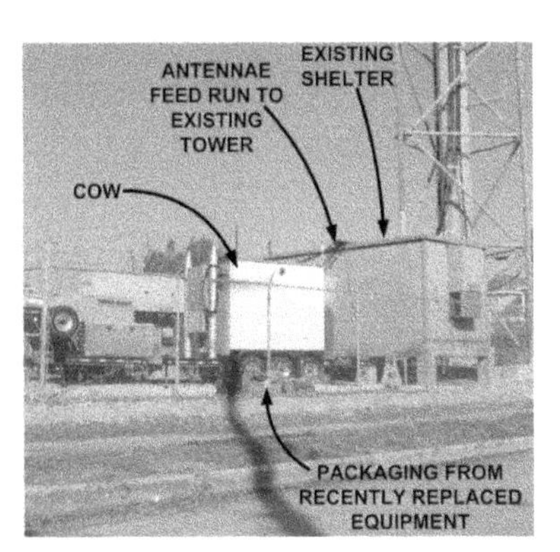

Figure 3.18 (a) Left: a COW. (b) Right: a satellite linked COLT

logistical needs typically increases for wireless network operators with respect to PSTN service providers because cell sites diesel gensets – both portable and permanent – usually require to be refueled every day. Due to the already discussed more flexible characteristics of wireless networks when compared to the PSTN, service restoration of damaged MTSOs also tends to be simpler than for damaged COs. When a MTSO is damaged, its functions, and base station and subscriber database are transferred to an undamaged MTSO. Yet, higher traffic in the latter switch may limit the number of calls in the area of the former switch.

3.3.3 Other Communication Systems

In general, other communication systems have the same restoration strategies that those discussed for the PSTN and wireless networks. However, it is relevant to mention some particular cases of interest. One concerns restoration of CATV network elements requiring grid's power, such as amplifiers. Usually these network elements are located every few hundred meters, e.g. 100 to 300 meters apart. The problem appears because these amplifiers are usually located on poles. Although this location is useful in cases, such as in coastal areas during hurricanes, because it prevent direct damage from storm surges [81], their higher position makes them difficult to power from portable gensets. As a result, many times the implemented solution is ad-hoc, such as the one in Figure 3.19a using off-the-shelf camping portable generators. Besides being unsafe, particularly for earthquake prone areas where aftershocks are certain

Figure 3.19 (a) Left: ad-hoc backup genset for CATV. (b) Right: Mobile microwave transmission unit (right tower) next to an existing cell site (left tower)

to occur, this solution is impractical because refueling is difficult. Another case of interest is the one related with link restoration when a main fiber optic cable is severed. The solution depends on the length of the affected segment. For short segments a microwave links with mobile transmission sites, such as the one in Figure 3.19b, is typically the chosen alternative. For long distances satellite links are used instead. A final case that is worth to mention is the restoration of damaged datacenters-restoration strategies for datacenters lacking power are the same as those implemented in COs. Three critical functions are usually involved with datacenters restoration: data operation, data routing and data storage. Thanks to the distributed nature of data networks, with adequate infrastructure in place these functions can usually be easily restored by transferring these functions to unaffected datacenters. Hence, data restoration, like congestion in the PSTN, are problems occurring during disasters that are not primarily related with infrastructure issues and will not be further discussed herein.

3.4 Communication Systems Technology Alternatives for Disaster Impact Mitigation

Both direct and indirect effects must be considered in order to adequately plan communication systems infrastructure in order to achieve networks that are resilient to disasters. Risk is a fundamental concept in infrastructure planning for disaster resiliencies because it provides an objective way of evaluating technologies that are technically effective and that have an adequate cost. Risk is usually defined as the product of the probability of a given disaster and its impact, measured in terms of costs. However, a better definition within the technology planning context discussed here, in which different engineering choices can be made, is

$$R = P_D\left(\min(I_D V, I_M)\right) \tag{3.11}$$

where P_D is the probability of the expected disaster to happen, I_M is maximum possible impact, I_D is the expected disaster impact, and V is the system under analysis vulnerability, i.e., an indication of how much more or less susceptible a network element is to sustain the expected impact ID based on the site design. Since no impact can exceed I_M, the minimum between I_M and $I_D V$ is onsidered in (3.11). Risk can be evaluated based on three phases when a disaster affects communications infrastructure: during the disaster, immediate disaster aftermath, and long term aftermath [85]. The challenge of the infrastructure technology planning is to make appropriate engineering choices in order to minimize overall risk.

3.4.1 Infrastructure Risk Minimization during the Disaster Phase

During this phase, disaster impact analysis is concerned with both direct damages and indirect effects, i.e., outages due to lack of power. There are many studies and theories developed particularly among civil engineers directed to reduce infrastructure damage caused by disasters. For this reason, infrastructure hardening design to minimize these damages is not discussed in detail here. However, it is relevant to mention some practical aspects usually observed in the field. Although it is true that buried cables are more resilient to most disasters than overhead cables, during hurricanes elevated infrastructure tends to perform better than ground-level infrastructure. The reason is that communication infrastructure is no longer almost excluively based on a cable plant as it was until the 1980s. With a prevalent use of electronic components in PSTN outside plant and in wireless networks base stations, elevated infrastructure prevents damage from storm surges and flood waters, which are the most damaging actions of hurricanes, even several miles inlad as it occurred with Tropical Storm Allison when it affected the city of Houston in 2001. However, locating all DLC RTs and wireless base stations on platforms is economically unfeasible, which makes a risk assessment evaluation particularly important in order to determine in which sites the use of elevated platforms is more effective. Economic issues might be one reason why construction practices, particularly for wireless base stations lacks uniformity, with base stations above the flood plane and others at ground level even at the same cell site. When locating a base station higher above ground it is important in order to avoid damage to place all infrastructure elements at a safe height. Examples of the result of not following this practice are observed in Figure 3.20. Locating SAIs higher on poles may also be a suitable solution in order to reduce the large number of these cabinets often damaged during hurricanes or floods. Moving infrastructure from ground level to higher positions may not be necessarily disadvantageous towards earthquakes as many times soil liquefaction, ground cracks, and landslides cause significant damage to buried or ground level infrastructure elements. Still, adequate design is needed in order to avoid damage due to intense shaking which is magnified on poles (Figure 3.21).

The idea of rising infrastructure that requires local power relates also with the problem of how to address arguably communication systems leading failure cause: lack of power. As (3.8) suggests, power supply availability can be improved both in normal conditions or during disasters by increasing batteries capacity,and/or reducing U_a. One relevant techniques that facilitate reducing U_a is to have a diverse power supply by installing local electricity generators. Since it is assumed that in case a site has a permanent diesel generator the locally stored fuel lasts long enough to avoid fuel starvation even during a disaster, then the analysis in this phase focuses on reducing the chances of having the local generator damaged, whereas the case of exhausting the local diesel storage is part of the issues considered later in Phase 2 (aftermath). The option of increasing the energy stored locally – usually from batteries – follows the traditional telecom power plant architecture, i.e., an energy system. This is the approach chosen by the US Federal Communication Commission (FCC) in its mandate following Hurricane Katrina. This order (07-107) was resisted by some communications operators and it was eventually suspended by the US Office of Management and Budget (OMB) – the OMB is an office within the US executive branch that oversees the activities of federal agencies, such as the FCC. Although increasing the battery capacity may lead to the

Figure 3.20 Two cell sites affected by Ike. (a) Top: The shelter was not damaged but the ac panel located lower was severely damaged. No permanent gensets were present at the site. (b) Bottom: The base station cabinets were not damaged but a portable genset needed to be deployed in order to replace the permanent genset that was out of service because of damage received to its propane tank located at ground level

Figure 3.21 Damage due to intense shaking from the 2010 earthquake in Chile at a pole mounted CATV UPS in which batteries and equipment seem to have forced the door to open. Notice also the severed drops caused by fallen debris from sorrounding buildings

desired level of availability during a disaster, this approach has still some drawbacks. One of those is added energy storage high cost, particularly for large sites. Another one is that this solution does not address the root cause of communication systems power supply weakness during disaster: lack of a diverse power source. Yet another weakness of this approach is that air conditioners are not powered from batteries. Thus, adding more batteries does not prevent air conditioners failure. This failure may lead then to communication system outage because of over-temperature even with the batteries still sufficiently charged. This situation may occur when, for example, the grid is out and the genset fails because ground shaking during an earthquake damaged it, or because water from extremely intense rains during a hurricane contaminated the diesel in the fuel tank. A final problem with batteries is their heavy weight that makes active network elements more vulnerable to sustain structural problems in CO buildings, or in distributed elements placed in elevated platforms. One option to batteries, particularly for distributed network elements, is to use fuel cells as backup power sources. These are typically proton exchange membrane fuel cells that are fed from hydrogen tubes. These fuel cells have usually longer autonomy than equivalent battery systems and they weight less, but after a disaster it may difficult to secure hydrogen tubes for many sites.

The alternative solution instead of increasing local energy storage is to have diverse locally generated power sources. As Figure 3.22 represents, this alternative implies making the communications site power plant a micro-grid-micro-grids are independently controlled electric networks powered by relatively low power generators placed very close to the load. Micro-grids may still add batteries for increased availability, but, with an adequate design, micro-grids may achieve high availability without energy storage. In micro-grids the local generating units are their primary source of power, and hence, they are either operating or in hot stand-by, i.e., active and connected to the system in idling operation, as opposed to cold stand-by in which generators are not connected nor operating until required to do so. The electric grid is, then, a secondary power source that complements the local power units. The most common options for local generation units – also called microsources or distributed generation (DG) sources – are photovoltaic (PV) modules, low-power wind generators, microturbines, fuel cells with reformers, and reciprocating engines. Since all these microsource technologies have availabilities at least an order of magnitude lower than that of power grids, it is essential that at least two diverse DG technologies requiring each different sources of energy from the other are utilized in order to reach high availability

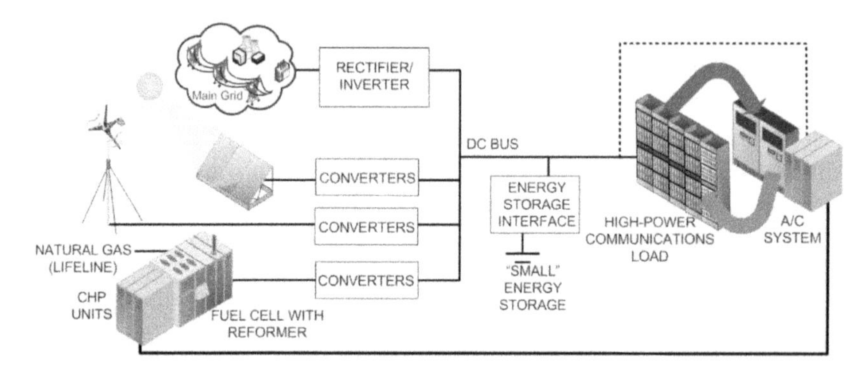

Figure 3.22 A micro-grid based telecom power and cooling infrastructure

levels. Each group of parallel connected DG sources of the same technology are called microsource clusters, Extensive calculations derived in [86] indicate that during normal operation and with an $n + 1$ arrangement of 2 source clusters it is possible to reach availabilities between 5 and 6 nines with minimal or no energy storage. An availability of 5-nines is obtained when a single converter – called center converter – is used for each microsource cluster, whereas 6 nines is achieved with redundant configuration of conventional single-input converter modules. Although center converters yield lower availabilities they are more economical than several conventional converter modules. A good compromise solution that allows reaching availabilities equal to that of conventional single-input converter modules but at a lower cost is to use multiple-input converter modules [86].

Micro-grids also provide an alternative to address the aforementioned availability issues related with air conditioning. Many microsource technologies allow integrating heating and cooling circuits, either through direct connection of the air conditioner to the dc bus, or through combined heat and power cycles (see Figure 3.22). Unfortunately, this last solution is currently only suitable for COs, datacenters, or other large sites. Hence, micro-grids based power plants avoid the inherent availability weakness in air conditioner electric supply observed in conventional communications energy systems architecture. Although nowadays microsources cost is a concern that makes micro-grid application in normal conditions to be cost-effective only when applied to medium and large sites [87], a risk assessment evaluation considering the cost of outages during disasters may yield good results for lower power sites as well [85]. Furthermore, micro-grids may significantly reduce

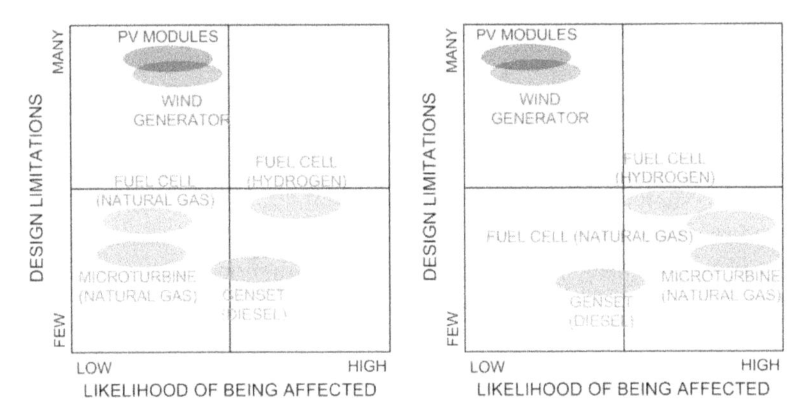

Figure 3.23 Relevant characteristics of microsources in Phase 1 of a disaster (a) Left: hurricanes. (b) Right: earthquakes

cost of batteries – which could reach as much as 40% in a typical conventional telecom power plant – because only a small amount of energy storage is needed for load following. Additional cost benefits can be achieved by increased efficiency and operation cost. These last benefits are achieved thanks to micro-grid independent control that allows for operating microsources and converters at their maximum efficiency point instead of depending on a non-optimal operation set by the power grid as observed in conventional telecom power plants.

Figure 3.23 shows a simplified description of the microsources characteristics in terms of the main concerns when an earthquake or a tropical cyclone happens: to avoid both outages caused by lack of power and damage. A critical aspect of microsources performance during the phase when the disaster is acting is whether or not lifelines are required, and if so how likely it is for these lifelines to be affected. For example, most microturbines, some reciprocating engines, and fuel cells with reformers typically consume natural gas. Natural gas is in general a good option for areas at risk of tropical cyclones but not for earthquake prone areas because its service is interrupted in case of a seism even when there is no observed damage to avoid fires. Some microsources, such as PV modules and small wind turbines do not need lifelines and with a good design they can survive hurricanes and earthquakes without damage. However, their generation profile is variable, e.g., PV systems cannot generate power during the night, so a reliable power supply will require either large energy storage or another microsource technology that can produce power when there is no wind or not enough solar radiation. Although PV modules and small wind generators can maintain operation during earthquakes they cannot operate during the several hours under the effects of a tropical cyclone. An additional concern with these microsources is their sizing as explained in more detail in the next phase. Although other alternatives, such as reciprocating engines and a few microturbines, that are powered from locally stored fuel, such as diesel, propane, or kerosene, do not require lifelines during the disaster, they need refueling during the immediate aftermath, when roads conditions may be an important concern.

3.4.2 Infrastructure Alternatives during the Immediate Aftermath Phase

In this phase, it is assumed that the disaster has passed, so direct effects leading to damage are no longer an issue. Thus, only indirect effects leading to power related outages, such as genset fuel starvation due to fuel delivery issues, influence decisions for this phase. Since the goal of the technology choices made for this phase is to minimize the risk associated with the impact of outages cost, the focus is on restoration work and repairs while avoiding loss of power supply. Therefore, logistical and repair operations and needs are fundamental aspects of the technology evaluation in this phase. Consequently, an efficient planning for smooth logistical operation performed before a disaster strikes may reduce risks in this phase, particularly from power losses. For example, in the US each wireless operator is usually responsible for its own power infrastructure at a shared cell site. As a result it is common to find multiple portable gensets deployed at each site. An adequate planning may, instead, lead to agreements so a single deployed genset is shared by all wireless operators at the site. This alternative significantly reduces the risk for logistical problems which the main issue at this phase without implying any difference in the probability of site failure for each network operator which would otherwise still have a single generator for each of its own base stations. Another aspect that will influence the technology choices for this phase is both how fast damaged infrastructure of each particular network operator and needed lifelines can be repaired after an expected disaster of a given intensity strikes. In terms of the worst case scenario regarding communications power the analysis typically focuses on how simple and fast it is possible to deploy and operate a mobile power plant. These and other restoration techniques have been previously discussed when describing past disasters, such as difficulties associated with portable gensets in outside plant application, and for

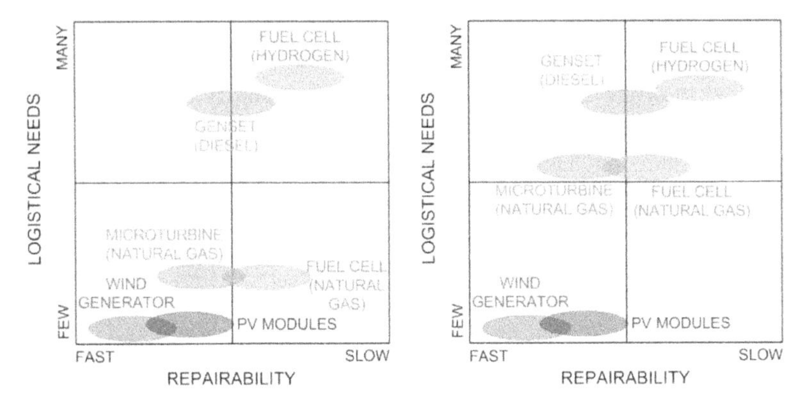

Figure 3.24 Relevant characteristics of microsources in Phase 1 of a disaster (a) Left: hurricanes. (b) Right: earthquakes

that reason they will not be further discussed in here. However, since power issues have been identified as an important source of outages during disasters, it is relevant to comment some of the potential effects that alternative power technologies may have during this phase. There are also a large number of technologies that can be used to restore service, such as using wireless local loop systems for PSTN outside plant circuit restoration. But since power issues are more critical, these other options will not be further discussed in this chapter.

Figure 3.24 represents the critical technological aspects considered when evaluating microsources during phase 2. Certainly, PV modules and small wind generators are the ones that have less logistical requirements because no lifelines are necessary provided that the system has enough batteries to compensate the power output when there is no wind or enough solar radiation. However, either by themselves or in hybrid systems, they have two problems: high cost and large footprint. Consider for example a typical base station located in the Mississippi River Delta, south of New Orleans, and that consumes 5 kW. In a day, its energy consumption is 120 kWh. Consider now that the goal is to power this base station with PV modules and batteries after a hurricane struck the area at the peak of the Atlantic hurricane season, in September. If it is assumed that the PV modules are placed facing south with a tilt equal to the local latitude (30°), the average total solar radiation is about 5 kWh/m^2/day [88]. Let us also consider that 200 W Sanyo's HIT Power 200 modules are chosen. However, in ideal weather conditions, these solar panels can generate its maximum output power during a few hours of the day. Hence, still with ideal weather conditions, additional capacity needs to be installed in order to compensate power deficits early in the morning or in the afternoon, and to charge the batteries to power the site during nighttime. These solar panels have an area a little over 1 m^2 and an efficiency of approximately 17%. Hence, the effective average solar radiation is approximately 0.85 $kWh/m^2/day$. Therefore, at least 140 PV modules are needed. In reality, it is desirable to add some more modules in order to compensate internal conversion losses in the power electronic interfaces, shadowing, and to provide redundancy. Thus, the final number of PV modules is about 164. Since each module measures 1.35×0.9 meters and they have a 30° tilt then the total area of all these PV modules is 146 m^2; obviously, impractical. However, some sites may be equipped with less PV capacity than the load. Although this solution may not avoid outages by itself, it may reduce the logistical needs for a diesel genset located at the site by bringing diesel consumption down. As mentioned, microsources fuelled by natural gas have minimal logistical needs. However, natural gas service will likely be interrupted during earthquakes. Technology alternatives relaying on locally stored fuel, such as hydrogen for backup fuel cells, or diesel or propane for permanent generators, have logistical need because they require refuelling. Hence, in these cases the option that provides the largest autonomy in order to maximize the refuelling period is typically the desired one, particularly for distributed network elements both in the PSTN and in wireless networks. Thus, propane fuelled systems are typically more suitable than the others because they have longer autonomy for the same volume of fuel.

3.4.3 Infrastructure Alternatives during the Long Term Aftermath Phase

Once communication network infrastructure and its lifelines have been restored to the same capacity levels existing before the disaster there is yet more potential effect that a disaster can have on a communications network: excess infrastructure because the disaster have destroyed important portions of the demand. This excess capacity implies a cost associated with unused capital that is depreciated in the network operators accounting books. Once again, the lack of hardwired connection for wireless network is an advantageous because capacity can be simply adjusted by appropriate base station controls or by removing electronic cards. In contrast, even with the added flexibility provided by DLCs, PSTN adjustment to a much lower demand is more complicated, particularly in the outside plant because high costs of buried infrastructure cannot be recovered and additional infrastructure network elements, such as poles and aerial cables, can only be removed and reutilized in another place at a very high cost. In COs and MTSOs most electronic equipment in modern power plants has modular designs, and its components can be relocated elsewhere. One exception is the installation cost, which is particularly high for cables. This cost cannot be recovered either. Hence, special attention needs to be paid to demand evolution after disasters and to infrastructure reconfiguration ease.

3.5 Conclusions

Past experience from many disasters indicate that a significant portion of communication networks outages after disasters originates in infrastructure issues, particularly in network elelements power supply. In most of these disasters direct damage to facilities tends to be, in general, less significant than power-related outages. Theoretical analysis based on Markov models verifies these past experiences. This analysis indicates that the relatively low availability of power grids creates a floor to communication sites availability that cannot be raised to adequate levels through diesel backup generators only. Calculations indicate that adequate availability levels can only be reached with the help of several hours of batteries' energy storage. However, air conditioners cannot be fed by batteries so the cooling infrastructure is still dependent on the insufficient availability set by the power grid and eventually a diesel generator. Hence, a common failure mode during disasters is components overheating caused by air conditioners power supply failure. Furthermore, since relevant disasters, such as tropical cyclones and earthquakes, typically affect large portions of the power grid, communication sites grid tie become a single-point of failure and, thus, a weak point in communication systems operation.

The discussion points out that communication systems are not self sufficient because many of its network elements depend on other infrastructures, such as roads, in order to maintain operation. These other needed infrastructures, and logistical needs become more prevalent in distributed network elements, such as wireless base stations and DLC RTs, which account for a significant portion of communication network outages in past disasters. Common alternatives, such as deploying portable diesel generators tend to have practical problems and add more strain to logistical operations. Other alternatives, such as using natural gas gensets, can only be used in some areas and as a response to disasters, such as tropical cyclones, that do not tend to affect significantly natural gas supply. A suitabler option is to use micro-grids because with diversed power sources, they can achieve power supply availability higher than that in conventional power plants even without the need for batteries. However, application to low-power sites is limited and microsources deployment planning need to consider the impact of disasters on DG sources lifelines. Still, micro-grids are essential components in future developments of advanced smart grids. It is expected that smart grid development will reduce communication systems outages by addressing the problem of power supply availability at its origin: the relatively low availability of conventional power grids. Microgrids will allow smart grids to implement a decentralized .power generation and control architecture that overcomes many weaknesses found in conventional power grids caused by their centralized control and power generation architecture. With a higher power availability provided by smart grids it is expected that power issues found in distributed communication network elements and in air conditioners will be reduced leading to an overall improvement of communications survivability during disasters.

4

Self-Organizing Cognitive Disaster Relief Networks

Nuno Pratas[1,2], Nicola Marchetti[1], Neeli Rashmi Prasad[1], Antonio Rodrigues[2] and Ramjee Prasad[1]

[1]*Center for TeleInFrastruktur, Aalborg University, Denmark; e-mail: {nup,nm,np}@es.aau.dk*
[2]*IT/IST, Technical University of Lisbon, Portugal*

4.1 Introduction

4.1.1 What Is a Disaster?

A disaster, according to Jacobs and Gaver [121], can be loosely defined as an event, which over a relatively short time period, causes a large number of casualties and/or infrastructure damage. An instance of the latter is the breakdown in the affected area of the infrastructure based communication systems. Disasters are usually unexpected and may come with minimal warning and with uncertain degree of importance and corresponding effects. Disasters may be caused by:

- *Nature* – earthquakes, fires, floods, hurricanes/typhoons, epidemics, or combinations thereof.
- *Mankind* – industrial accidents (such as chemical, nuclear plant leakages), weapons of mass destruction (chemical, biological, or nuclear), attacks on cities, military installations, conceivably within buildings, but also on port, in-flight aircraft, multiple automobile accidents or aircraft crashes.

N. Marchetti (ed.), Telecommunications in Disaster Areas, 95–125.
© 2010 *River Publishers. All rights reserved.*

4.1.2 What Is a Disaster Relief Network?

To give relief in the affected area the interested parties, i.e. relief and security groups, need communications support, and therefore a communication network needs to be established as quickly and as easily as possible as to ensure that rescue and relief efforts are not further hindered. From a communications perspective, a disaster scenario can be characterized as having an hazardous spatial and possibly radio environments. A consequence of the first is the prevention of infrastructure deployment, due to time and economical constraints, while a consequence of the latter is the impairment of the spectrum bands assigned to disaster relief services. This impairment can be the result of jamming, which can be intentional or simply the result of un-organized deployment of disaster relief networks. As such, disaster relief and emergency scenarios share common characteristics such as the lack of infrastructure and are hazardous spatial and radio environment limited. These characteristics raise the need for specifically tailored wireless networks which are able to work in such scenarios. Further, the acknowledgment that infrastructure-based networks in such deployment areas may be destroyed raises the need to find alternatives based on infrastructure-less and decentralized wireless technologies [139].

Self-Organizing Cognitive Radio Networks come as a solution to implement disaster relief networks, these being infrastructure-less or not. But what are Self-Organizing Cognitive Radio Networks? To define these first we must explain what is Cognitive Radio, and finally what is a self-organizing network.

4.1.3 What Is Cognitive Radio and How Did It Appear?

The Cognitive Radio (CR) concept was first put forward by Mitola III in [119], as the natural evolution of the Software Defined Radio (SDR). The CR by Mitola was presented as an intelligent agent able to track radio resources and related computer-to-computer communications and able to detect user communications needs as a function of use context, and to provide radio resources and the wireless services most appropriate to those needs. Later the focus of research in CR was directed mainly towards the intelligent and opportunistic use of the radio resources, a technique also known as Dynamic Spectrum Access (DSA).

The Cognitive Radio concept was further refined in [107], where the DSA was emphasized, and a new definition emerged:

> an intelligent wireless communication system, aware of its surrounding environment (i.e., outside world), that uses the methodology of understanding-by-building to learn from the environment and to adapt its internal states to statistical variations in the incoming RF stimuli by making corresponding changes in certain operating parameters (e.g., transmit-power, carrier-frequency, and modulation strategy) in real-time, with two primary objectives in mind: highly reliable communications whenever and wherever needed and efficient use of the radio spectrum.

As stated in the definition, the CR is a device that is aware (of its surrounding radio environment), is intelligent (as it decides the best approach to convey the wireless service), learns and adapts, is reliable and above all efficient. One of the main capabilities of the CR, is its ability to reconfigure, which is enabled by the SDR platform, upon which the CR is built.

4.1.4 Why Do We Need Cognitive Radio?

The rapid growth of a multitude of wireless communication services had as a consequence the increase on the demand for electromagnetic radio frequency spectrum, which is a scarce resource, mainly assigned to license holders on a long-term basis for large geographical regions, causing, according to measurements [93], a large portion of the spectrum to remain unused for a significant percentage of the time. Upon this scenario, the Federal Communications Commission in the US and the European Commission's Radio Spectrum Policy Group in the EU, proposed a secondary and concurrent usage of this spectrum, focusing on the case where this secondary system does not interfere with the normal operation of the license holders. These new regulations, in contrast with the licensed bands to which entities such as TV stations or cellular operators are granted exclusive access, allows the evolution of a new paradigm, consisting of devices with the ability to adapt to their spectral environment and able to make use of the available spectrum in an opportunistic manner, i.e. performing Dynamic Spectrum Access.

4.1.5 Why Is It Important to Be Used in Disaster Relief Networks?

Disaster Relief Networks need to be reliable, so as not to hamper the performance of rescue and relief missions. Since the spectrum available to these networks can become crowded leading to un-intentional jamming, especially in cases where due to the dimension of the disaster there are several relief teams, and using uncoordinated communication systems, leading to situations where the systems may interfere with each other, decreasing the performance and potentially making them unusable. Through the use of Self-Organizing Cognitive Radio Networks it is possible through the DSA to work outside that crowded spectrum if the spectrum policies are updated to allow for it.

4.1.6 Why Should It Be Self-Organizing, i.e., How Does the Cognitive Radio Paradigm Enable This?

A disaster relief network to be usable and robust needs to intertwine the self-organizing/autonomic and cognitive radio paradigms. To understand why a disaster relief network needs to be self-organizing, we first need to understand what is the self-organizing/autonomic paradigm and why it is emerging as a key technology requirement for next generation networks.

The self-organizing/autonomic networking concept, follows the concept of Autonomic Computing, introduced by IBM [112], and where the aim is to create self-managing networks able to the overcome the rapidly growing complexity of current and future networks. An autonomic system as such, is one that is self-configuring, self-optimizing, self-healing and self-protecting, therefore requiring minimal intervention, mostly related to policy adjustments. The motivation for developing self-organizing networks is to minimize the cost of deploying and running a network, i.e. the Capital Expenditures (CAPEX) and Operation Expenditures (OPEX). The reduction of the CAPEX can be achieved by reducing the need of manual intervention to configure the equipment at the time of deployment, while the reduction of the OPEX is obtained by reducing or eliminating the need of manual intervention to maintain and reconfigure the network during its lifecycle. The ultimate goal is therefore to reduce the cost and retail price of data services.

In this chapter we present how the Cognitive Radio and Autonomic Networking paradigms can be combined to form a solution for a disaster relief network. We start by discussing the Cognitive Radio fundamentals and giving an overview of the components of a Cognitive Radio network. Then the autonomic networking fundamentals are discussed. Afterwards we present deployment scenarios over which the added functionalities of this kind of network are demonstrated to be useful. The chapter ends with a wrap-up of the presented concepts.

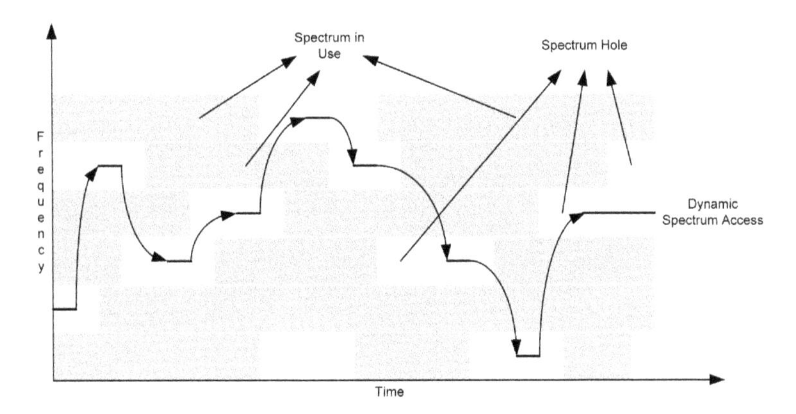

Figure 4.1 Spectrum hole concept [107]

4.2 Cognitive Radio Fundamentals

The key enabling technologies of Cognitive Radio are the functions that provide the capability to share the spectrum in an opportunistic manner. In [95] a summarized definition of CR was presented: "A Cognitive Radio is a radio that can change its transmitter parameters based on interaction with the environment in which it operates."

Since most of the spectrum is already assigned, the challenge is to share the spectrum with coexisting networks without interfering with their transmission. For this the cognitive radio enables the usage of temporarily unused spectrum, which is referred in the literature as Spectrum Hole or white space, which is depicted in Figure 4.1. In [107], the definition of spectrum hole was given: "A spectrum hole is a band of frequencies assigned to a primary user, but, at a particular time and specific geographic location, the band is not being used by that user." If this spectrum hole where the CR is operating starts also to be used by another secondary user, then the CR moves to another spectrum hole or stays in the same, altering its transmission power level or modulation scheme to avoid interference. Therefore the Cognitive Radio is a paradigm for the efficient and opportunistic use of the available spectrum and through it, it is possible to:

- Determine which portions of the spectrum are available;
- Select the best available channel;
- Coordinate access to this channel with other users;
- Vacate it when the channel conditions worsen.

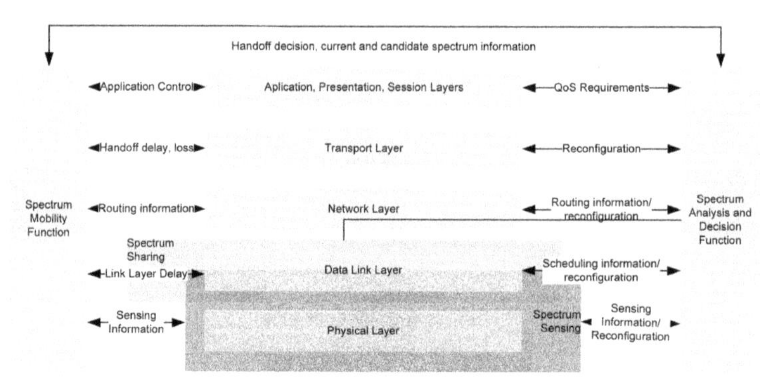

Figure 4.2 model updated with the Cognitive Radio functionalities [95]

To implement the Cognitive Radio paradigm a network needs to employ adaptive network protocols. Such an example is given in [95, 110], where it is proposed an adaptation of the OSI network model to account for the Cognitive Radio paradigm, and where a cross layer approach is used. We depict in Figure 4.2 the model proposed in [95], that is used as a basis in this chapter. In this model, the spectrum sensing and spectrum sharing functions cooperate with each other to improve the network spectrum use efficiency. In the spectrum decision and spectrum mobility functions, application, transport, routing, medium access and physical layer functionalities are carried out in a cooperative way, considering the dynamic nature of the underlying spectrum.

The OSI model added functionalities which implement the Cognitive Radio paradigm and their purpose are explained in the next sections.

4.3 Spectrum Sensing

The Spectrum Sensing function monitors the state of the spectrum environment at the network node position, with the purpose of detecting unused spectrum, i.e. the spectrum holes. In Figure 4.3 we depict the spectrum sensing as an imperfect and simplified mapping of the radio environment to a representation in the sensing node. By sensing the spectrum the Cognitive Radio becomes aware of and sensitive to the changes in its surrounding, giving to the Cognitive Radio the information needed to adapt to its environment.

Spectrum sensing, covered extensively in the literature [95,126], falls into the domain of the detection theory, presented in detail in [135]. Spectrum sensing is realized as a physical and MAC layer mechanism. The physical

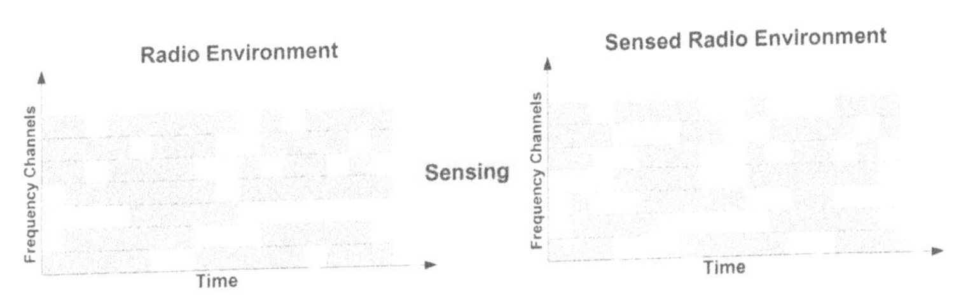

Figure 4.3 Representation of sensing from a cognitive radio prespective [123]

layer sensing focuses on detecting signals, and the detection methods put in place can be classified into two groups, either coherent (prior information needed, e.g. Pilot Detection [95]) or non-coherent (no prior information needed, e.g. Energy Detector [134]). The MAC layer part of the spectrum sensing focuses on when to sense (in time) and which spectrum to sense (in frequency).

The performance of the sensing depends on the local channel conditions, i.e. depend on the multipath, shadowing and local interference. The conjunction of these conditions can result in regimes where the signal SNR is below the detection threshold of the sensor, resulting in missed detections or false alarms creating the imperfect mapping illustrated in Figure 4.3. To overcome this limitation there have been several proposals in the literature such as in [106, 118, 136], where it was proposed the use of cooperation. Since the signal strength varies with the sensor location, the worst fading conditions can be avoided if multiple sensors in different spatial locations share their local sensing measurements, i.e. take advantage of the spatial diversity. Most of these proposed cooperative methods are based on data fusion techniques to perform the decision on what is the actual state of the spectrum. In [114, 123, 124], the use of scheduling schemes to select which channels to sense based on the channel statistics is proposed.

4.4 Spectrum Analysis and Decision

The CR purpose is to enable networks to use the appropriate available spectrum band according to the Quality of Service requirements. To accomplish this, new spectrum management functions are required, taking into consideration the dynamic spectrum characteristics. These functions are spectrum sensing, spectrum analysis, and spectrum decision, and their interrelation is depicted in Figure 4.4. While spectrum sensing is primarily a PHY and MAC

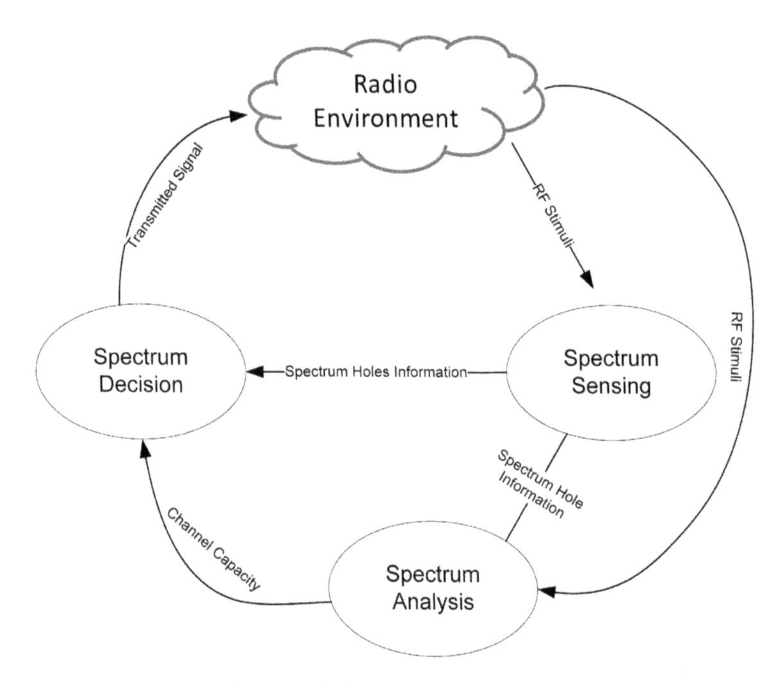

Figure 4.4 The cognitive cycle [107]

layer issue, spectrum analysis and spectrum decision are closely related to the upper layers.

It is expected that the spectrum sensing function will find Spectrum Holes (SH) spread over a wide frequency range including both unlicensed and licensed bands. Therefore, these will potentially show different characteristics according not only to the time varying radio environment but also to the spectrum band information such as the operating frequency and the bandwidth.

Due to the dynamic nature of the underlying spectrum the communication protocols need to adapt to the wireless channel parameters, since the behavior of each protocol affects the performance of all the other protocols built on top of it. For example, by using different medium access techniques CR networks, the round trip time (RTT) for the transport protocols will be affected. Similarly, when re-routing is done because of link failures arising from spectrum mobility, the RTT and error probability in the communication path will change accordingly. The change in error probability also affects the performance of the medium access protocols. Consequently, all these changes affect the overall quality of the user applications.

The Spectrum Analysis and Decision function cooperates with the communication layers, as shown in Figure 4.2. In order to decide the appropriate spectrum band, the information regarding the QoS requirement, transport, routing, scheduling, and sensing is required. So the use of a cross layer approach will use the interdependencies among functionalities of the communication stack, and their close coupling with the physical layer to accomplish the Analysis and Decision function.

4.5 Spectrum Analysis

The SH identified by the spectrum sensing function have different characteristics which vary over time. The purpose of the Spectrum Analysis function is to characterize these spectrum bands, as to identify the appropriate one for the node requirements.

To account for the dynamic nature of networks, each SH should also consider not only the time-varying radio environment, but also the interferers activity and the spectrum band information such as operating frequency and bandwidth. Hence, it is essential to define characterizing parameters that can represent the quality of a particular spectrum band. The following were identified in [95]:

- Interference – The spectrum band in use determines the characteristics of the interferers affecting the channel.
- Path loss – The path loss increases as the operating frequency increases. If the transmission power of a node remains constant, then its transmission range decreases at higher frequencies.
- Link errors – Depending on the modulation scheme and the interference level of the spectrum band, the error rate of the channel changes.
- Link layer delay – Depending of the spectrum band in use, it is expected different path loss, wireless link error, and interference. All these conditions amount to different link layer packet transmission delay.
- Link layer delay – Depending of the spectrum band in use, it is expected different path loss, wireless link error, and interference. All these conditions amount to different link layer packet transmission delay.
- Holding time – The activities of interferers can affect the channel quality to the network. Holding time refers to the expected time duration that the node can occupy a band before getting interrupted. The lower is the

holding time the higher is the frequency of spectrum handoff. Since the spectrum handoff also means wasting time in adjusting the transmission to a new channel, then the system throughput and connectivity are sacrificed in these procedures. So channels with longer holding times are therefore better. Since frequent spectrum handoff can decrease the holding time, previous statistical patterns of handoff can be considered.

The spectrum band is characterized by the channel capacity, which can be derived from the above parameters. The SNR is normally used to perform channel capacity estimation, but since it only considers the observations at the receiver, then the previous parameters need also to be considered to estimate the channel capacity.

4.6 Spectrum Decision

Upon characterization of the available spectrum bands and the associated potential estimated channel capacity, the appropriate operating spectrum band can be selected. Based on the node QoS requirements, the data rate, acceptable error rate, delay bound, transmission mode, and bandwidth can be determined. Then according to the decision rule in use, the set of appropriate spectrum bands can then be chosen.

Several examples of rules to be used in the Spectrum Decision function can be found in the literature. Here we briefly highlight some of them. In [142], five spectrum decision rules are presented, which are focused on fairness and communication cost, however assuming that all channels have similar throughput capacity. In [111], an opportunistic frequency channel skipping protocol is proposed for the search of channels with better quality, and where the decision is based on the channel SNR. In [94], an adaptive based centralized decision solution is presented, which also considers spectrum sharing. The adaptation mechanism considers the user traffic and the base stations hardware resources.

4.7 Spectrum Mobility

The purpose of the Spectrum Mobility function is to allow a network to use the spectrum in a dynamic manner, i.e. allowing the CR nodes to operate in the best available frequency band. The Spectrum Mobility function is defined as the process through which a node changes its frequency of operation, also known as spectrum handoff [95].

In a CR network, the spectrum mobility arises when the conditions of the channel in use by the node become worse, due to the node movement or because an interferer appears in the channel. The Spectrum Mobility gives rise to a new type of handoff, referred to as spectrum handoff in [95]. A CR can adapt to the frequency of operation. Therefore, each time a CR node changes its frequency of operation, the network protocols are going to shift from one mode of operation to another. The protocols for different layers of the network stack therefore need to adapt to the channel transmission parameters of the operating frequency, as well as being transparent to the spectrum handoff and the associated latency.

The purpose of spectrum mobility management in CR networks is to make sure that such transitions are made smoothly and as soon as possible such that the applications running on a CR node perceive minimum performance degradation during a spectrum handoff. It is therefore essential for the mobility management protocols to learn in advance about the duration of a spectrum handoff. This information can be provided by the sensing algorithm, through the estimation of the channel holding time. Once the mobility management protocols learn about this latency, their job is to make sure that the ongoing communications of a CR node undergo only minimum performance degradation. Whenever a spectrum handoff occurs, there is an increase in latency, which directly affects the performance of the communication protocols. Thus, the main challenge in spectrum mobility is to reduce the latency for spectrum handoff which is correlated to the spectrum sensing latency. During spectrum handoff, the channel parameters such as path loss, interference, wireless link error rate, and link layer delay are influenced by the dynamic use of the spectrum. On the other hand, the changes in the PHY and MAC channel parameters can initiate spectrum handoff. Moreover, the user application may request spectrum handoff to find a better quality spectrum band.

As shown in Figure 4.2, the Spectrum Mobility function cooperates with Spectrum Decision and Analysis function and Spectrum Sensing to decide on an available spectrum band. In order to estimate the effect of the spectrum handoff latency, information about the link layer and sensing delays are required. Moreover, the transport and application layer need to be aware of the latency to reduce the abrupt quality degradation. In addition, the routing information is also important for the route recovery algorithms which base their decisions also on the information estimated about the frequency of spectrum handoff on each of the available links. For these reasons, the spectrum mobility is closely related to the operations in all communication layers.

4.8 Spectrum Sharing

The shared nature of the wireless channel requires the coordination of transmission attempts between CR users. Spectrum sharing can be regarded to be similar to generic MAC problems in traditional systems. However, substantially different challenges exist for spectrum sharing in CR networks. The coexistence with other systems and the wide range of available spectrum are the main reasons for these unique challenges.

In [95] an overview of the steps of spectrum sharing in CR networks was provided. The spectrum sharing process consists of five steps:

- Spectrum sensing – When a CR node aims to transmit packets, it first needs to be aware of the spectrum usage around its vicinity.
- Spectrum allocation – Based on the spectrum availability, the node can then allocate a channel. This allocation does not only depend on spectrum availability, but it is also determined based on policies.
- Spectrum access – Since there may be multiple CR nodes trying to access the spectrum, this access should be coordinated in order to prevent multiple users colliding in overlapping portions of the spectrum.
- Transmitter-receiver handshake – Once a portion of the spectrum is determined for communication, the receiver should also be informed about the selected spectrum.
- Spectrum mobility – When the conditions of the allocated spectrum deteriorate, the CR nodes need to move to another vacant portion of the spectrum, making use of the spectrum mobility function.

The existing work in the literature regarding spectrum sharing can be classified in three aspects, being those architecture, spectrum allocation behavior and spectrum access technique.

The classification of spectrum sharing techniques based on the architecture is as follows:

- Centralized – A centralized entity controls the spectrum allocation and access procedures. To aid the procedures, a distributed sensing procedure is proposed such that each entity in the network forwards its measurements about the spectrum allocation to the central entity and this entity then constructs a spectrum allocation map. Examples of this kind of architecture can be found in [94, 99, 128, 130–132, 138].
- Distributed – Distributed solutions are mainly proposed for cases where the construction of an infrastructure is not preferable. Each node is responsible for the spectrum allocation and access is based on local or

global use policies. These policies can be vendor specific or can be dictated by an regulator entity, like the FCC. An example can be found in [109].

The classification of spectrum sharing techniques based on the access behavior is as follows:

- Cooperative – Cooperative solutions consider the effect of the node's communications on other nodes. The interference measurements of each node are shared with other nodes, and the spectrum allocation algorithms also consider this information. All centralized solutions are regarded as cooperative, although there are also distributed cooperative solutions. Examples of these can be found in [99, 109, 110, 115, 132, 140].
- Non-cooperative – Non-cooperative solutions consider only the node at hand. These solutions are also referred to as selfish. While non-cooperative solutions may result in reduced spectrum utilization, they do not require the exchange of control information among other nodes as the cooperative ones do. Examples of these can be found in [122, 141, 142].

When comparing these approaches in terms of architecture and access behavior, it was shown in the literature that cooperative approaches outperform non-cooperative ones, moreover it was shown that distributed solutions closely follow centralized solutions. Evidence of these results can be found in [123, 124, 141].

The classification of spectrum sharing techniques based on the access technology is as follows:

- Overlay – In overlay spectrum sharing, a node accesses the network using a portion of the spectrum that is not used by licensed users. As a result, interference to the primary system is minimized [99, 104, 110, 120, 122, 141, 143].
- Underlay – Underlay spectrum sharing exploits the spread spectrum techniques developed for cellular networks, an example can be found in [109]. Once a spectrum allocation map has been acquired, a CR node transmits in a way such that its transmitting power at a certain portion of the spectrum is regarded as noise by the licensed users. This technique requires sophisticated spread spectrum techniques and can use increased bandwidth when compared to overlay techniques.

The theoretical work on spectrum access in CR networks reveals important tradeoffs for the design of spectrum access protocols. It was shown that cooperative settings result in higher utilization of the spectrum as well as improved fairness. However, this advantage may eventually not be so high considering the cost of cooperation due to the signaling overhead. In [104, 117] it was shown that the spectrum access technique, i.e., whether it is overlay or underlay, always affects the performance of legacy systems. While an overlay technique focuses on the holes in the spectrum, dynamic spreading techniques are required for underlay techniques for interference free operation between concurrent systems.

The performance of spectrum sharing directly depends on the spectrum sensing capabilities of the CR nodes. Spectrum sensing is primarily a PHY layer function when considering the detection. However, in the case of cooperative detection it encompasses also the MAC layer, since the CR nodes need to make use of a common channel for exchanging sensing information. It is clear that the performance of communication protocols depend on spectrum sensing, i.e. on getting accurate information about the spectrum utilization at the CR nodes locations. This therefore implies a cross-layer design between spectrum sharing and spectrum sensing.

4.9 Physical Architecture and Re-configurability

The CR needs to be implemented on top of a hardware platform which enables its functionality. In Figure 4.5 a generic architecture of a CR transceiver is depicted, based on the one proposed in [110]. The main components of a CR transceiver are the radio front-end and the baseband processing unit. Each component can be reconfigured via a control bus to adapt to the time-varying RF environment. In the RF front-end, the received signal is amplified, mixed and A/D converted. The baseband processing unit of a CR is essentially similar to existing transceivers. The solution to enable this is the use of Software Defined Radio platforms, an example of which currently available in the market is the hardware enabling the GNU radio software stack [89,90].

The functions mentioned before, i.e. spectrum sensing, analysis and decision, enable the cognitive capability, provide the spectrum awareness, whereas the re-configurability enables the radio to be dynamically programmed according to the radio environment. More specifically, the re-configurability is the capability of adjusting operation parameters for the transmission on-the-fly without any modifications on the hardware components.

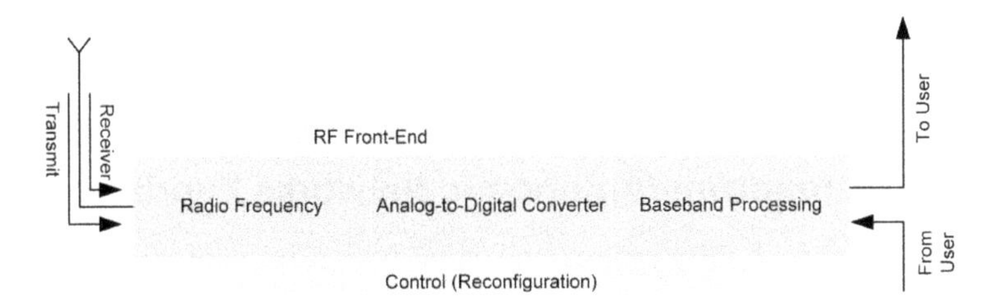

Figure 4.5 Cognitive radio transceiver

This capability enables the CR to easily adapt to the dynamic radio environment. In [95] one enunciates what should be the main reconfigurable parameters to be implemented in the CR, those being:

- Operating frequency – Based on the radio environment information, the most suitable operating frequency can be determined, enabling the communication to be dynamically performed in the appropriate frequency.
- Modulation – A CR node should reconfigure the modulation scheme in a way that is adaptive to the user requirements and channel conditions, i.e., in the case of delay sensitive applications, the data rate is more important than the error rate. Thus, the modulation scheme that enables the higher spectral efficiency should be selected. Conversely, the loss-sensitive applications necessitate modulation schemes with low bit error rate.
- Transmission power – Power control enables dynamic transmission power configuration within the permissible power limit. If higher power operation is not necessary, the CR reduces the transmitter power to a level that allows more users to share the spectrum, decreasing the interference.
- Communication technology – A CR can also be used to provide interoperability among different communication systems, therefore employing the SDR capabilities.

The transmission parameters of a CR can be reconfigured not only at the beginning, but also during a transmission. According to the spectrum characteristics, these parameters can be reconfigured such that the CR is switched to a different spectrum band, the transmitter and receiver parameters are reconfigured and the appropriate communication protocol parameters and modulation schemes are used. A re-configuration framework has been proposed in [108].

4.10 Self-Organizing/Autonomic Networks Fundamentals

The phenomenon of self-organization is pervasive in many areas of life. In nature, e.g., fish organize themselves to swim in well structured swarms, ants find shortest routes to food sources, and fireflies emit light flashes in a synchronized fashion. Other examples of self-organized behavior can be observed in economy, population dynamics, psychology, and brain theory. In all the above examples, the participating entities establish an organizational structure that does not require any central coordination. Instead, the entities interact directly with each other, reacting to changes in their local environment. Typically, such self-organizing systems are very flexible, adaptive, failure-robust, and scalable [125].

Communication network complexity has been an ever-increasing problem to deal with for engineers over the decades, and recent advances in self-organization indicate that this is an area to go for solutions. The Internet continues to grow unabated as the unifying technology, especially with the rapid adoption and convergence of mobile, fixed wireless, and broadband wireline access, connecting a myriad of mobile terminals, devices, and sensors at homes and businesses. At the same time, applications span body, personal, home, vehicle, and wide area networks. All this is adding to the spatio-temporal complexity of network topology and dynamics, increasing the burden on network administrators and users. This leads to the question of how this increased complexity can be reasonably managed without requiring users to become technical experts in the field, and network owners to spend time and resources in managing networks. In short, a fresh look is needed to design and develop networks that self-organize themselves or minimize human intervention as much as possible.

Lately, with the availability and analysis of volumes of traces and statistics on the behavior of nodes, two major approaches to describe complex networks have been proposed, related to the small world and scale-free concepts. Small world is a property observed in many systems, such as in society and communication networks, and refers to the fact that entities (e.g. people or network nodes) are closer than one may think. Most entities are connected such that the average path length between them is small, the entities tend to form clusters, and the connectivity distribution peaks at an average value, and then decays exponentially. Scale-free properties are those which do not

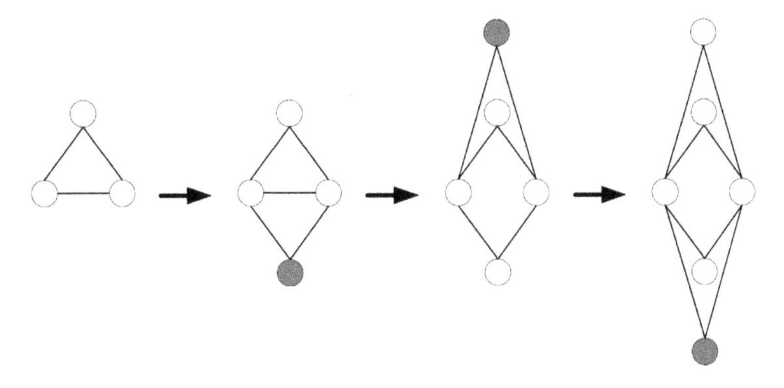

Figure 4.6 Growth and Preferential Attachment proprieties [116]

change with the size of the system, that in our case is the network. A particular scale-free property is that the connectivity distributions follow the power law form. This means that the connectivity distribution in a scale-free network tends to remain the same irrespectively of the size of the network, even at different orders of magnitude. Simply stated, the average path length between any two entities does not increase, independently of how large the network grows [102, 103]; this property is observed in many complex networks (e.g. the Internet). For Self-Organizing Networks (SON) applied to wireless communications, the fundamental scale-free and small world properties are also set by the limited radio propagation range [116]. In [97, 137] a scale-free network model was developed using the following properties observed in real networks:

- Growth – Real networks expand continuously by the addition of new nodes.
- Preferential attachment – New nodes attach preferentially to nodes that are already well connected.

The properties of growth and preferential attachment lead to the scale-free property and are typically exhibited by real large networks, and this is confirmed by observations revealing that in large networks most of the nodes have very few connections, while only a few nodes, called hubs, have many connections [137]. This phenomenon is depicted in Figure 4.6.

The application of SON concepts to wireless networks has gained momentum recently; indeed SON are seen a key driver for improving Operation and Maintenance (OAM) as they can help to reduce the cost of installation and management by simplifying operational tasks through automated mech-

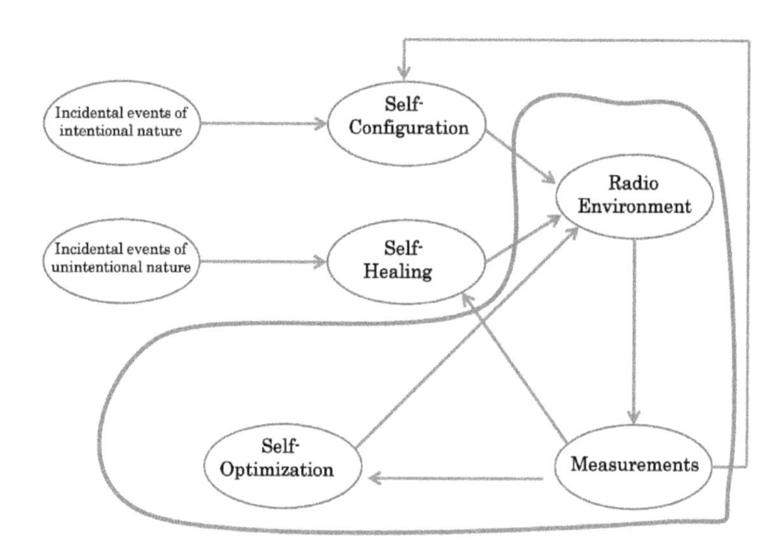

Figure 4.7 Self-Organizing Network cycle [116]

anisms targeting self-configuration, self-optimization and self-healing, which constitute the three main phases of the SON cycle, depicted in Figure 4.7.

4.11 Rationale behind SON

SON are likely to have a relevant impact on Capital Expenditure (CAPEX) and Operational Expenditure (OPEX). Indeed, currently about 17% of wireless operators CAPEX is spent on engineering and installation services [100], therefore the self-configuring functions can eliminate the on-site operations for basic settings and updates, which will reduce the CAPEX. Also, since about 24% of a typical wireless operator revenue goes to network OPEX, the self-optimizing functions of SON can help to reduce the workload associated with site survey and network performance analysis. Moreover, the energy saving functions derived by the application of SON will reduce the costs in terms of power consumed by the equipment. On top of this, self-optimization and self-healing will help to improve the user perceived quality, by optimizing network parameters and by mitigating quality degradations. However, the main rationale for wireless SON is not the current OPEX situation, but what is probable to happen next. In summary, for SON applied to wireless networks, the main driver is the outlook of increased cost, primarily OPEX, for emerging networks.

To exemplify the explanation of the purpose of the phases depicted in Figure 4.7, we consider the case of a cellular network in the following:

- Self-configuration – The self-configuration phase is triggered by incidental events of an intentional nature, e.g. the addition of a new site or the introduction of a new service or network feature. These upgrades generally require an initial (re)configuration of a number of radio parameters or resource management algorithms, e.g. pilot powers and neighbors' lists. These parameters have to be set prior to operation and before they can be optimized as part of the continuous self-optimization process [113].
- Self-optimization – In the self-optimization phase, intelligent methods apply to the processed measurements to derive an updated set of radio (resource management) parameters, including e.g. antenna parameters (tilt, azimuth), power settings (including pilot, control and traffic channels), neighbors' lists (cell IDs and associated weights), and other radio resource management parameters (admission/congestion/handover control and packet scheduling) [113].
- Self-healing – Triggered by incidental events of a non-intentional nature, such as the failure of a cell or site, self-healing methods aim to resolve the loss of coverage or capacity induced by such events up to the extent possible. This is done by appropriately adjusting the parameters and algorithms in surrounding cells. Once the actual failure has been repaired, all parameters are restored to their original settings [137].

4.12 Self-Organizing Cognitive Networks

Let us consider a CR network (CN) formed by spatially distributed nodes, where the nodes cooperate with each other through local interactions, and are able to adapt their states in response to both data collected locally as well as to data received from the surrounding nodes. Considering the Self-Organizing Paradigm, then when information arrives at any particular node, it creates a ripple effect that propagates throughout the network by means of a diffusion process. This results in a form of collective intelligence leading to improved adaptation, learning, tracking, and convergence behavior with respect to non-cognitive networks.

The edges linking the nodes can be assigned adjustable weights in accordance with the quality of the information that is exchanged over these edges; in this way, CN can adjust their topologies as well. Distributed processing techniques over such adaptive networks do not experience one of the main drawbacks of classical centralized fusion methods, i.e. the fact that central fusion approaches limit the autonomy of the network and add a critical point of failure due to the presence of a central node [92].

It has been observed in social and biological sciences studies on animal flocking behavior that while each individual agent in an animal colony is not capable of complex behavior, the combined coordination among multiple agents leads instead to the manifestation of regular patterns of behavior and swarm intelligence. In a similar manner, CN should benefit from local cooperation among the nodes, leading to enhanced performance in terms of e.g. improved learning, robustness, and convergence abilities [92].

CN can be designed to perform a variety of tasks such as detection, estimation, or resource allocation, through distributed processing. CN applications include environmental monitoring, distributed event detection, resource monitoring, target tracking, cooperation among CR searching for spectral resources, among others [92].

The cognitive cycle followed by CR, depicted in Figure 4.4, in their quest for free spectrum can be seen as a special case of a subset of the SON cycle showed in Figure 4.7 (the one circled at the right-bottom of that figure), i.e. a special case of the self-optimization phase of a SON cycle. As a matter of fact, the ubiquitous and pervasive computing and networking is becoming a clear trend in wireless communications: this is confirmed by the fact that computer processors are becoming part of more and more everyday items, which may form a wireless network. This trend corresponds to a shift from the very large to the very small, leading to electronics with the following characteristics: low-cost, miniature size and self-contained from energy perspective [127]. The ultimate goal is to achieve reliable universal coverage at all times and this brings to several challenges [129]:

- The large amount of devices will lead to rapidly run out of spectrum.
- Most devices have and will have limitations in energy consumption.
- Wireless might be unreliable when compared with the wired counterpart.
- The very high heterogeneity of devices might lead to incompatibility.

The above challenges in such complex networks' scenarios can be faced only if a certain degree of self-organization is put into the system. In particular, self-optimized/cognitive networks will play a crucial role in identifying the needed spectrum of the proper quality for the several coexisting systems, devices and applications. Examples of possible scenarios for self-optimized and, in particular, cognitive networks, are [98]:

- Vehicular networking – among cars, e.g. targeting accidents warnings, and within the car, e.g. establishing a network among personal devices.
- Wireless sensor networks and wearable computing – applied e.g. to monitoring of humans, in the context of assisted living and medical engineering.
- Disaster relief networks – among emergency services, to ensure that any time all the actors are able to communicate.

4.13 Deployment Scenarios

In this section we describe the four possible network architectures for a Disaster Relief Network: Cellular, Ad-hoc, Mesh and Satellite. In the following we will describe the strengths and weaknesses of these architectures from a disaster relief perspective.

4.13.1 Cellular Architecture

A communication system based on a Cellular network architecture uses a large number of wireless transmitters to create cells, as depicted in Figure 4.8. The cells are the basic geographic service area of a wireless communications system. By varying the downlink power levels is possible to size the cells according to the subscribers density and demand within a particular region. As mobile users travel from cell to cell, their conversations are *handed off* between cells in order to maintain a seamless service. Channels used in one cell can be reused in another cell that is situated some distance away, depending on the reuse factor. Cells can be added to accommodate growth, creating new cells in un-served areas or overlaying cells in existing areas [101, 133].

The advantages of the Cellular Architecture in general and from a Disaster Relief Network perspective are:

- Robust geographical coverage area – Each point-to-multipoint network (or sector) within a cellular network is considered to be one cell. A cellular network with multiple cells can cover a larger geographical area than a terrestrial single point-to-multipoint network. In the case the disaster area increases, the network coverage can be extended by adding additional base stations. The backbone that supports these base stations can be wired or wireless.
- Network capacity scalability – Each additional cell in a cellular network increases the number of end users that the network can serve. For example, a cellular network with two cells can serve twice as many end

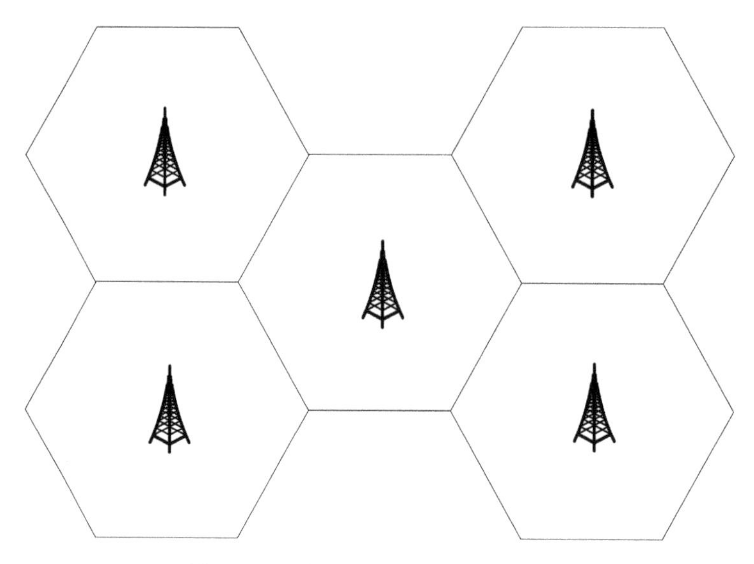

Figure 4.8 Cellular network architecture

users as a single point-to-multipoint network [101]. In disaster relief scenario this is important because more relief teams might join the effort and therefore more resources are needed.

- Network resource management – A cellular network allows the sharing of network resources between large numbers of end users. These resources include backbone bandwidth network administration, maintenance, and monitoring. From a disaster relief standpoint, this enables a better management of the available radio resources, especially when compared to decentralized solutions, like the ad-hoc networks.
- End user coverage redundancy – Cells in a cellular network might be designed to overlap with each other. If cells do overlap, end users might be able to obtain service from an adjacent cell if one cell is temporarily out of service. This especially important in disaster relief ensures that all users are always covered by at least one cell.
- Allows roaming – Roaming refers to the capability of an end user to move out of the coverage area of one cell and into the coverage area of another cell while retaining network connectivity.
- Reduced interference – Considering that all the cells in the cellular network are working in the same frequency range, then by using a frequency reuse schemes, power control, and proper deployment planning

it is possible to control and therefore reduce the interference. From a disaster relief point of view, this ensures that the network works optimally, but since it requires deployment planning, it might not be possible to accomplish, due to the time constraints associated with the relief operations.

The disadvantages of the Cellular Architecture:

- Backbone is connected to the base station – The primary disadvantage of a cellular architecture, is the need for a backbone to support it, i.e., to interconnect the base stations. In a disaster relief scenario there might not be a possible to setup a backbone link to connect all the base stations. This backbone can be wired or wireless, being the later preferred in disaster relief networks due to the increased ease of installation, although the wireless backbone can be subject to interference, decreasing the performance of the network.
- Power supply – The power supply needed for base stations is an important disadvantage, since normally they cannot be supported for a long time with a battery. There are of course alternative power supply solutions like portable generators fueled by diesel, or renewable energy sources like solar or wind.
- The base station – The base station is a focal node in the cellular architecture, and if it fails all the user nodes in that cell lose connectivity to the network. This is the common drawback of using a centralized network solution. Adjacent cells' base stations can take the user nodes from the defunct cell, but that goes with an increase of the load of those cells, reducing the amount of resources available to those same users when the base station was active.

4.14 Ad-Hoc and Mesh Architecture

The definition of an ad-hoc network has been formulated by the Internet Engineering Task Force (IETF) work group on Mobile Ad-hoc Networks (MANET) [91]:

> A mobile ad hoc network is an autonomous system of mobile routers connected by wireless links – the union of which form an arbitrary graph. The routers are free to move randomly and organize themselves arbitrarily; thus, the network's wireless topology may

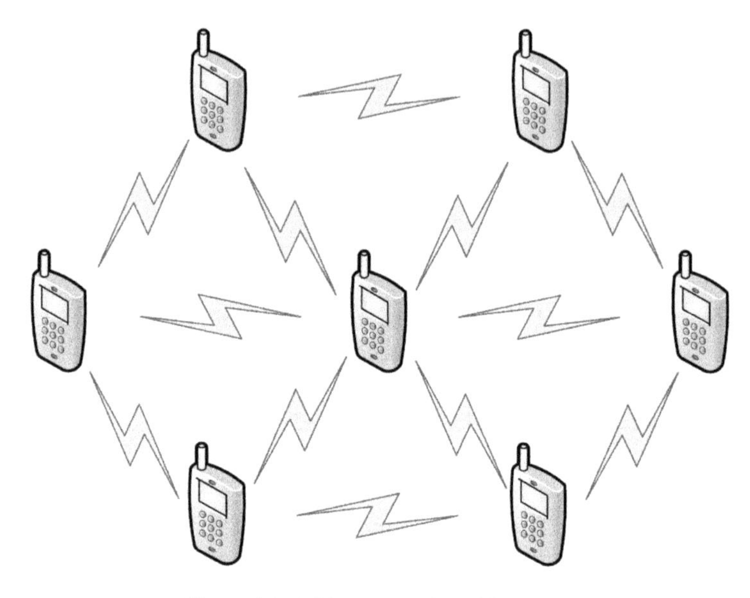

Figure 4.9 Ad-hoc network architecture

change rapidly and unpredictably. Such a network may operate in a standalone fashion, or may be connected to the larger Internet.

In Figure 4.9 is illustrated an example of an ad-hoc network architecture.

The mesh architecture is a step forward with respect to the Ad-hoc architecture, since it enhances the decentralized architecture by introducing network nodes with functionality traditionally only available in cellular networks, e.g., the base stations work both as a central nodes as well as ad-hoc nodes. The mesh architecture is a multipoint-to-multipoint architecture with one or more interconnection points. In Figure 4.10 is depicted a mesh network architecture. In a mesh network, each network node can connect to any other network node that is turned on and within wireless range. Mesh networks are usually deployed in areas where many end users are located relatively close to each other, such as from one block up to one mile apart. Each mesh network node performs two functions: as a wireless router/repeater or as an end node [96, 133].

Wireless mesh networks (WMNs) consist of mesh routers and mesh clients, where mesh routers have minimal mobility and form the backbone of WMNs. They provide network access for both mesh and conventional clients. The integration of WMNs with other networks such as the Internet, cellular,

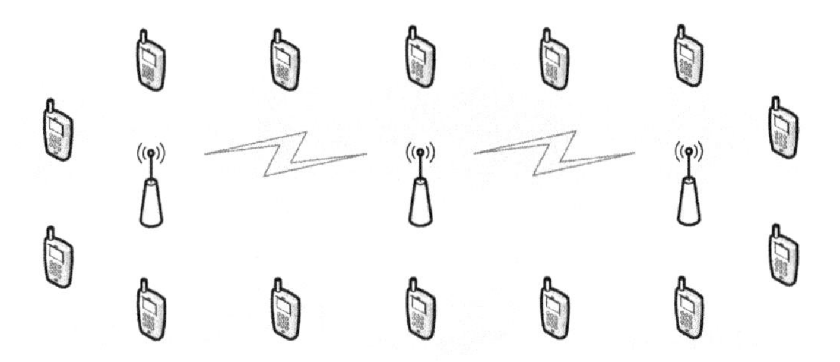

Figure 4.10 Mesh network architecture

IEEE 802.11, IEEE 802.15, IEEE 802.16, sensor networks, etc., can be accomplished through the gateway and bridging functions in the mesh routers. Mesh clients can be either stationary or mobile, and can form a client mesh network among themselves and with mesh routers. WMNs are anticipated to resolve the limitations and to significantly improve the performance of ad hoc networks, wireless local area networks (WLANs), wireless personal area networks (WPANs), and wireless metropolitan area networks (WMANs). They are undergoing rapid progress and inspiring numerous deployments. WMNs will deliver wireless services for a large variety of applications in personal, local, campus, and metropolitan areas. Despite recent advances in wireless mesh networking, many research challenges remain in all protocol layers [96, 133].

The advantages of mesh architecture are:

- Near-line-of-sight coverage – A clear Line of Sight (LOS) microwave path is always desirable between nodes in a wireless network. In many areas, clear LOS paths are difficult to achieve due to obstructions such as trees and buildings. In areas where LOS paths are partially obstructed, mesh architecture might allow reliable connectivity. This occurs because the network nodes are placed close enough together to allow reliable node-to-node communication in spite of the obstructions attenuating the signal strength. The denser the obstructions, the higher the signal attenuation and the closer together the mesh nodes need to be placed. This is important in disaster relief operations occurring in urban environments, where LOS conditions not always possible.
- Routing redundancy – In a mesh network, every wireless node is a wireless router or repeater. Operating as a router, each node is cap-

able of dynamically calculating the best available path to each distant node. If one node is down (due to failure, being turned off, or being blocked by an obstruction), network traffic is rerouted through other nearby nodes. This on-the-fly rerouting capability provides a measure of network routing redundancy, therefore ensuring network connectivity to all the network nodes.

- Simpler network design process – Point-to-multipoint and cellular network architectures require a moderately complex network design process to ensure that LOS paths exist between each end user location and the wireless access point. The network design process for a mesh architecture network is somewhat simpler. The node-to-node-to-node capability of a mesh network, where each node serves as a relay point, simplifies the network design process. As long as the network designer ensures that each node can communicate with at least one or two nearby nodes, network design can proceed without needing to verify the existence of longer LOS paths. This makes a mesh network a perfect solution for a disaster relief network.

The disadvantages of the mesh and ad-hoc architectures are:

- Higher density of wireless nodes – In a mesh architecture, wireless paths are shorter than in other network architectures. This means that if the number of network nodes is equal, the coverage area of a mesh network is less than the coverage area of a point-to-multipoint network, e.g. a cellular network. Therefore, limiting the possibility to deploy a disaster relief network, with a mesh architecture, over a large area.
- Progressive network deployment process – A mesh network relies on the availability of point-to-point-to-point wireless repeating. This means that if a wireless node is installed beyond the range of the nodes that have already been installed, the new node will not be able to connect to the mesh network to obtain service. To prevent this problem, one must deploy a mesh network progressively within a particular service area. Deployment must begin near the Internet access point and continue outward toward the far edge of the mesh.
- Difficult bandwidth management – Each end user node in a mesh network has dual roles: wireless router and end user node. The wireless bandwidth available to each end user node is shared between providing Internet access for that specific end user and providing backbone access for the other end users who are connecting to the internet through that node. The more hops exist between two nodes, the less bandwidth is

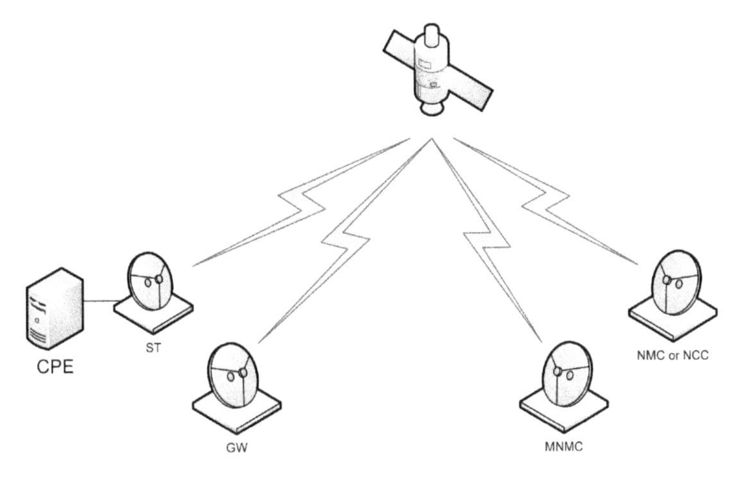

Figure 4.11 Satellite network architecture

available to that end user. This occurs because the relay nodes have also to relay data between the other nodes in the network.

4.15 Satellite Architecture

Satellite networks typically provide two types of service: broadcast and two-way communication services. In Figure 4.11 the architecture of a satellite network is illustrated.

A single satellite can deploy one or more satellite networks. A satellite network relies on a ground segment and utilizes some satellite on-board resources. The ground segment is composed of a user segment and a control and management segment [105].

In the user segment, Satellite Terminals (ST) connected to the end-user Customer Premises Equipment (CPE), directly or through a LAN and hub or gateway stations, connected to terrestrial networks [105]:

- ST are earth stations connected to CPE, sending carriers to or receiving carriers from a satellite. They constitute the satellite access points of a network.
- CPE are also called User Terminals (UT) and they include equipment such as telephone sets, television sets and personal computers. User terminals are independent of network technology and can be used for terrestrial as well as satellite networks.

- The gateway earth station (GW) provides internetworking functions between the satellite network and the terrestrial network, wired or wireless.

The control and management segment consists of [105]:

- Mission and Network Management Centre (MNMC) in charge of non-real-time, high-level management functions for all the satellite networks that are deployed in the coverage of a satellite.
- Network Management Centers (NMC), also called interactive network management centers, for non-real-time management functions related to a single satellite network.
- Network Control Centers (NCC) for real-time control of the connections and associated resources allocated to terminals that constitute one satellite network.

A satellite network comprises a set of ST, one or more gateways and one NCC that is operated by one operator and uses a subset of the satellite resources. Satellite networks are characterized by their topology (meshed, star or multi-star), the types of link they support and the connectivity they offer between the earth stations:

- Meshed network topology – In a meshed network, as defined previously, every node is able to communicate with every other node. A meshed satellite network consists of a set of earth stations which can communicate one with another by means of satellite links. This entails that several carriers can simultaneously access a given satellite and subsets of these carriers simultaneously access a given receiver. A meshed satellite network can rely on a transparent or a regenerative satellite. In the case of a transparent satellite, the radio-frequency link quality between any two earth stations in the network must be high enough to provide the end users with a service achieving the target bit error rate. In the case of a regenerative satellite, the on-board demodulation of the signal puts fewer constraints on the signal power received from the earth stations.
- Star network topology – In a star network, each node can communicate only with a single central node, often called the hub. In a multi-star topology, several central nodes are identified. The other nodes can communicate only with those central nodes.

Two types of link can be established through a satellite network: unidirectional links, where one or several stations only transmit and other earth stations only receive, and bidirectional links, where earth stations both transmit and receive. Unidirectional links are usually associated with a star topology, in satellite broadcast-oriented networks. Bidirectional links can be associated with a star or meshed topology and are required to transport any two-way telecommunication services, e.g. internet and telephony services.

The advantages of satellite network are the following:

- Global coverage – Satellite communications can be available virtually everywhere on earth, through the use a constellation of satellites. Giving coverage to disaster relief operations occurring in remote places.
- Higher coverage footprint per satellite – The coverage of a single satellite is far more extensive than what any terrestrial network can achieve.
- Easy to serve remote areas – Provides a telecommunications infrastructure to areas where terrestrial alternatives are unavailable, unreliable or simply too expensive.
- Broadcast/asymmetric focused – Satellite's inherent strength is as a broadcast medium making it ideal for the simultaneous distribution of bandwidth-intensive information to hundreds or thousands of locations.
- Higher bandwidth – Due to higher coverage footprint, satellite networks can cover quickly hundreds to thousands of locations, connecting cities or remote locations across a large landmass, where copper or fiber cost is prohibitive.
- Scalability – Satellite networks are scalable, allowing users to expand their communications networks and their available bandwidth, by contracting more bandwidth or satellite time, and by adding new satellites or constellation of satellites.
- Complementing solution – Satellites can be integrated to complement, extend a terrestrial communications network, helping to overcome geographical barriers, terrestrial network limitations and other constraining infrastructure issues. This is especially useful for disaster relief networks, where a terrestrial backbone might not be possible to be established.

The disadvantages of the Satellite Networks Architecture are:

- Setup cost is high – When compared with ground networks, the cost of setting up (building, putting it in orbit) a geostationary satellite or a constellation of satellites, is many orders of magnitude higher.
- Complexity of the transceivers – The satellite terrestrial transceivers need an higher complexity than the transceivers used in terrestrial networks, to compensate for the channel characteristics of downlink

and uplink satellite communications, e.g., the very high path loss, the Doppler effect when considering non geostationary satellites.

- Higher delay – The distances between the terrestrial transceivers and the satellite are much higher than the terrestrial networks; in the case of a geostationary satellite, the propagation delay may vary from 238 to 278 ms, this depending on the position of the transceiver relatively to the satellite. While in terrestrial network the propagation delay is below 1 ms.
- Limited indoor coverage – The attenuation in indoor environments is too high therefore the indoor coverage cannot be achieved directly by the satellite, most likely needing a transceiver dish with line of sight to allow for it.

4.16 Disaster Relief Network Architecture

The architecture of a disaster relief network is the most important feature enabling the network itself to be robust. So considering the previously listed architectures, the best option to achieve a robust network is to combine all the architectures, so to ensure that in the case where one of the control elements of one of the architectures breaks down, then by changing the network architecture it is possible to maintain its connectivity. This dynamic changing of the network architecture is only possible through the implementation of a self-organizing algorithm. An example of this is illustrated in Figure 4.12, where the backbone of the network is provided by a satellite link.

The most balanced architecture, considering the advantages and disadvantages, is the Mesh Network, since it combines the strength of a Cellular Network with the robustness of an Ad-hoc Network. The use of a satellite backbone in this case only makes sense if the network needs a backbone, i.e., if the network is deployed over a large area, or if a terrestrial backbone cannot be deployed.

4.17 Conclusion

In this chapter we discussed the motivations to build disaster relief communication systems which implement the Cognitive Radio and Self-Organizing paradigms, as well presenting the possible architectures on which these paradigms can be implemented.

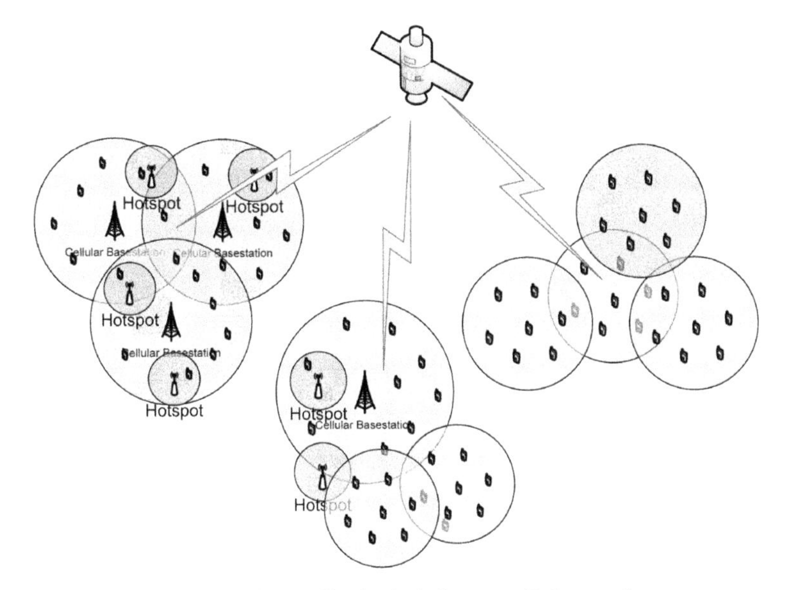

Figure 4.12 Satellite backed disaster relief network

We described the meaning of each of these paradigms, and why they are important to the disaster relief scenarios. Namely, the Self Organizing paradigm allows the network to self-optimize enabling it be efficient and robust, even in the case where the architecture changes. The Cognitive Radio paradigm enables the network to strive to have always available radio resources, through the use of dynamic spectrum access.

As a final note, this kind of network is possible to be implemented using the results of the ongoing research in the field, but it still needs the regulating bodies to adapt the spectrum allocation rules, so that during emergency situations, such networks can make use of the dynamic spectrum access.

5

Modeling Public Safety Scenarios for Emergency Communication Systems

Nils Aschenbruck

Institute of Computer Science 4, University of Bonn,
Roemerstr. 164, 53117 Bonn, Germany;
e-mail: aschenbruck@cs.uni-bonn.de

5.1 Introduction

As emergency communication systems need to be as reliable as possible, the performance of these systems has to be evaluated thoroughly. Field-tests are probably the evaluation method preferred. However, they are quite expensive, as sufficient number of devices is needed. Furthermore, the results concerning scalability and reproducibility are limited. Thus, especially for the evaluation of algorithms and protocols, computer simulations are an alternative.

Naturally, the results of such simulative performance evaluation strongly depend on the models used. In the last few years, a lot of research and also (simulative) performance analysis for multi-hop networks (such Mobile Ad-hoc Networks (MANETs), mesh networks, and sensor networks) has been done. These networks are often motivated by public safety scenarios which scenarios require communication systems as robust as possible. As multi-hop networks do not rely on infrastructure, they meet this requirement by their very definition.

In multi-hop networks, instead of using infrastructure, each node may act as a potential relay depending on its current position and the traffic patterns. Thus, the movements of the nodes, the traffic, and the distribution of the traffic to the nodes do have a significant impact on the results of the simulative performance evaluation. Therefore, realistic models are needed.

N. Marchetti (ed.), Telecommunications in Disaster Areas, 127–146.
© 2010 *River Publishers. All rights reserved.*

Various mobility models were proposed during last decade. There have been several general surveys [145, 151, 152, 155] as well as some specific ones for vehicular models [163, 167]. However, there are only a few approaches that model the specific characteristics for public safety scenarios (see Section 5.2). In Section 5.3, we will introduce these models.

For the evaluation of network performance aspects, modeling the data traffic realistically is an important issue. In MANET simulation studies, uniformly or exponentially distributed constant bit rate traffic is widely used (e.g. [154, 156, 159, 171, 175, 188]). Some approaches (e.g. [162]) simulate voice traffic based on studies in telecommunication or cellular systems, others (e.g. [161, 183, 187]) use file transfer protocol (FTP) traffic for performance evaluation. There are several traffic models. However, often constant length as well as a constant number of traffic streams are assumed. Furthermore, the distribution of the traffic sources is important as well. Especially for infrastructure-less multi-hop networks, the distribution of the traffic sources influences the results of the performance evaluation. Often, uniform distributions are assumed, but public safety scenarios are typically structured in a tactical manner. In disaster area scenarios nodes do not move randomly. Thus, a uniform distribution cannot be assumed. In Section 5.4, we will introduce the traffic and traffic distribution models available for public safety scenarios.

A more specific and detailed model makes only sense if the included details have an impact on the results of a performance evaluation with the model. In Section 5.5, we will show that a detailed modeling of public safety scenarios has an impact on performance evaluations. Finally, we will provide a brief summary of the chapter (Section 5.6).

5.2 Public Safety Scenarios

Public safety scenarios are often classified into *normal events* and *disasters*. Scenarios that can be handled by the regular day-by-day public safety units (e.g. firefighters and first responders) are seen as normal scenarios. As soon as the regular units need additional help, such as civil protection or a larger number of non-regular or far-away units, the scenario is seen as a disaster. In detail, the scenarios may be classified according to three criteria (cf. [177]):

- Environment: indoor, urban, rural;
- Coverage: single spot, wide area;
- Situation: day-by-day routine, emergency, disaster.

Of course, there are dependencies between the different categories. If an emergency reaches a certain coverage, it will most probably be seen as a disaster. Depending on the number of day-by-day units in an environment, a scenario may have to be categorized as a disaster as well. In general, a disaster is defined as a situation in which the number of normal day-by-day units is not sufficient.

Furthermore, the classification depends on the tactical situation which contains:

- Kind and size of the incident;
- Expected development;
- Number of people affected (especially lost and injured ones);
- Specific dangers such as chemical, biological, nuclear ones;
- Time of the scenario;
- Complexity of the event.

Depending on the tactical situation different numbers of specific units are needed.

Concerning movements and data traffic of the units specific characteristics can be observed independently of the actual scenario.

5.2.1 Movements

In all public safety scenarios, movements and actions are strictly structured. Units do not walk around randomly, but have a goal and follow orders which are based on tactical reasons. These tactics often influence the speed of a moving unit. The destinations depend on the working site which is based on tactical issues. The tactics as well as the scene are usually hierarchically organized. Typically, the site is divided into different tactical areas. Each unit belongs to one of these areas. The area in which a unit moves depends on tactical issues and is often restricted to one specific tactical area. For example, in a disaster, a firefighter belongs to an *incident site* and a paramedic will work at one place in the *casualties treatment area*. First responders are assigned to one station and do their service typically around their station. Different kind of units also move with different velocities. An ambulance moves faster than a pedestrian fire-fighter. Of course, all units look for optimal paths that help them to fulfill their tasks in an efficient way.

Beside tactical issues there are geographic restrictions that influence the movements. If it is an urban or indoor scenario, there are walls and streets that restrict the movements to specific paths. Even in a disaster where everything might have been destroyed, there are obstacles that have an influence on the movements.

Furthermore, movements in public safety scenarios are normally movement in groups. There are teams that work together. Thus, there are spatial dependencies to other persons nearby. For example, in a disaster area scenario there are often four people carrying an injured person. In ambulances there are typically two or three people.

However, for the performance evaluation only individuals that have a communication device are of interest. Often only one individual of such a tactical formation has a communication device. Thus, from the communication perspective, there is no group movement. However, in the future, it may be affordable or necessary to provide each individual with its own communication device.

As a conclusion, there are the following main characteristics for movements in public safety scenarios:

- Hierarchical organization;
- Tactical areas;
- Optimal paths;
- Heterogeneous velocity;
- Geographic restrictions;
- Group movement (optional);

5.2.2 Data Traffic

Today, the main application in public safety scenarios is voice communication: different users communicate via push to talk voice calls. The users that communicate with each other (*talk group*) share one broadcast voice communication channel. Technically this broadcast voice communication channel may be realized, e.g. as a separated physical channel or as a multicast group. The term *talk group* does not restrict the technical realization.

For future networks the following applications are expected for public safety scenarios:

- Voice: as voice communication has been used for years, it will be also used in the future;
- Information Systems: there will be information systems to share data and enable better command and control as well as situational awareness;
- Automatic Vehicle Tracking: especially for dispatched units such as ambulances the tracking of the units is important;

- E-Mail: may be used for simple message transfer as well as command and control;
- Pictures: as cameras are omnipresent in current mobile devices, there can be pictures to be transmitted;
- Videos: may be used in addition to voice communication for different purposes such as remote diagnosis;
- Sensor data: by using specialized sensors it is possible to provide specific data, such as wind, temperature, and contamination.

Concerning the modeling there are two challenges. First, it is hard to know the impact of each application for future networks. Second, if an application does not exist, it is impossible to do an accurate modeling.

Thus, it is a good idea to start with modeling the voice communication that exists today. In the future, this voice communication may evolve: the voice codecs may change, the technical realization of a talk group may change, or the voice traffic may become video traffic (video-phone). However, the communication time will still be dominated by the actual message to be said. The message to be said (holding time), the pause between two messages (idle time), as well as the one who speaks will not change. Considering the distribution of traffic sinks, each member of a talk group is a sink of his traffic group (as long as his device is switched on).

Due to the hierarchical structure in public safety scenarios there are typically both local and global talk groups. Local talk groups communicate inside a specific tactical area. Global talk groups are used for communication between different areas. The assignment of the nodes to a specific tactical area influences the distribution of traffic sources and sinks.

In general, each call of a talk group is done by one sender that starts speaking and stops after a certain amount of time. There is only a half-duplex connection (unlike a telephone call): while one user speaks, the others have to listen. Different calls with semantic connection (e.g. question and answer) may be regarded as one conversation or session, where a conversation consists of an arbitrary number of calls between two callers, and, typically, the callers alternate in calling each other. However, in each session there are only two users that communicate directly. Nevertheless, all the other members of the talk group need to listen to the whole communication for tactical reasons.

Furthermore, the communication pattern typically found in such public safety talk groups is *star communication*. There is one user that is the head of this talk group. Each user is only allowed to communicate with this head and

Table 5.1 Classification of public safety mobility model

Model		Environ.	Coverage	Situation	Scenario
Hostage Rescue	[170]	rural	single spot	emergency	Military – Hostage Rescue
Platoon	[178]	urban	single spot	emergency	Military – Platoon Movement
TIMM	[160]	indoor	single spot	day-by-day	Military – Urban Warfare
Disaster Area	[146]	rural	single spot	disaster	Civil – Catastrophe at Event
CORPS	[168]	indoor	single spot	day-by-day	Civil – Tunnel Rescue

not allowed to communicate directly among themselves. This pattern is used to manage the flood of information.

5.3 Modeling Mobility

In this section, we will introduce the mobility models that can deal with specific characteristics for public safety scenarios. Table 5.1 classifies the models available based on criteria described in Section 5.2. Furthermore, the scenario the model was created for is listed. The following subsections will describe these models in detail.

5.3.1 Hostage Rescue

In [170] a hostage rescue scenario was modeled. The movements of a squad are modeled in a quite detailed way. The scenario is separated into different phases: such as march, deployment, access, and fallback. In this scenario typical characteristics of public safety scenarios such as hierarchical organization and tactical areas are realized.

5.3.2 Platoon Mobility Model

In [178] two scenarios are presented. Both model the movements of a platoon in an urban area. The modeling is based on [144]. In the first scenario, the platoon is divided while moving into three squads. The squads take different paths on their way through the city. On the way several fireteams leave the squad temporary. At the end, all units merge back to one platoon and leave the urban area. In the second scenario, the platoon moves as whole. While moving, the platoon redeploys due to a more dangerous area and enemy contact.

In [179] the scenarios are extended to the *Coalition* mobility model. This model is a combination of Reference Point Group Mobility model (RPGM) [166] with a graph-based model [184]. There is a movement graph defined. The platoon moves on the graph and changes its formation at predefined waypoints.

These scenarios realize specific characteristics for military movements. Typical characteristics for public safety scenarios like hierarchical organization, tactical areas, optimal paths, geographic restrictions, and group movements are realized. However, the modeling is quite scenario specific as the paths on the movement graphs and the changes of the platoon's formation are predefined.

5.3.3 Tactical Indoor Mobility Model

The Tactical Indoor Mobility Model (TIMM) [160] models groups of soldiers in an urban warfare scenario. The goal of the soldiers is to secure a building. In the model the basic tactics of this kind of scenario are realized. The basic idea of one of these tactics is to divide all available units into small groups which enter the building consecutively. Each group secures a small part of the building, in such a way, that a fallback path is always secured. If necessary, all units can use this path to leave the building quickly. The groups start their movement one after another and stop if a parametrized distance has been traveled. Furthermore, a graph representation of the building is used to restrict the movements to valid paths. Rooms and doors are represented as vertexes while edges represent valid paths between the vertexes. Each time a group reaches its destination (a vertex in the graph), a new destination is chosen. Candidates are all neighboring vertexes from the current position that have not already been visited. If there is only one destination, the group takes this as the target destination. Otherwise, as long as the group is large enough, the group is segmented into smaller groups reaching the different possible destinations. If there is no neighboring non-visited vertex, the complete graph is searched for the nearest non-visited vertex. This is chosen as the next destination of the group. If there is no such vertex, all rooms have already been searched and the mission objective has been fulfilled. Thus, the model terminates.

In [160] specific characteristics of this model are examined. Furthermore, the model is validated based on a detailed scenario description. It is shown that the mobility model shows specific characteristics concerning the number of neighbors and number of link changes compared to standard models. Furthermore, there is a significant impact on simulative performance evaluations.

5.3.4 The Disaster Area Model

To realize area-based movement, the simulation area is divided into polygonal tactical areas. The tactical areas are classified according to tactical concepts. For some areas there are both stationary nodes, which stay in the respective area moving according to a random based mobility model, as well as transport nodes that carry the patients to the next area following a movement cycle. Different areas and classes allow heterogeneous velocities. The area and the class (stationary or transport) the node belongs to define the movement of the node as well as the minimal and maximal speed distinguishing pedestrians from vehicles.

The optimal path for the movement of the transport units between the different areas is determined by methods of robot motion planning. For finding the shortest paths and avoiding obstacles between the tactical areas, visibility graphs are used. A visibility graph is a graph where its vertices are the vertices of the polygons. There is an edge between two vertices, if the vertices can "see" each other – meaning the edge does not intersect the interior of any obstacle. The shortest path between two points consists of an appropriate subset of the edges of the visibility graph. Thus, after having calculated the visibility graph containing all possible shortest paths between the areas avoiding obstacles, the direct path between two areas for each transport unit can be calculated.

Vehicular transport units (e.g. ambulances) typically leave the disaster area to carry patients to hospital. Thus, joining and leaving nodes are realized using specific entry and exit points (registration areas).

Group mobility is realized as an optional characteristic for disaster areas, as in civil protection there may only be one device for each group. Nevertheless, it is realized similarly to reference point group mobility model (RPGM) [166]. The units of each area are grouped. The size of the group depends on the type of the area and of the group. Similarly to RPGM the nodes follow their reference point. The movement of each node in a group is calculated in relation to the movement of the reference point.

In [146] specific characteristics of the disaster area model are examined and compared to other less detailed models. The disaster area model shows a specific node distribution that results in heterogeneous node density and larger relative mobility. Furthermore, there is worse connectivity and even for larger communication ranges links tend to break quicker. These specific characteristics do also have an impact on the results of performance evaluation. If one uses the disaster area model for evaluating protocols, the results will change drastically compared to standard models.

5.3.5 CORPS

In [168] an event-driven mobility model for first responders at an incident site is described. The model is called Cooperation, Organization, and Responsiveness in Public Safety (CORPS). A typical scenario for this model is a tunnel rescue operation. The model consists of three parts. The first part (first responder model) defines coverage area, roles, and groups. The second part (event model) defines time and location of incidents. The third part (interaction model) defines the mobility patterns based on the other two parts. Depending on a node's role and the type of an event specific movements are triggered. To realize shortest paths an algorithm similar to the one based on visibility graphs that is also used in the disaster area model is implemented.

The model realizes specific characteristics for public safety scenarios. Optimal paths and geographic restrictions are modeled in a quite detailed way. The model allows to generate different kind of scenarios in an automatic way. However, the model is quite complex and especially a parametrization of a realistic complex scenario is challenging.

5.4 Modeling Traffic

Traffic analysis and modeling for public safety communication systems has been done for several years. At first (in the 1980s), group communication in land mobile radio systems was analyzed to derive models to design and deploy the new trunked radio systems (e.g. [164,165,182]). Similar to earlier studies of telephone voice conversations (e.g. [153,169]), session lengths and interarrival times were found to be exponentially distributed. They follow a 2-state Markov model assuming one state as talk spurt (*ON-state*) and the other as silence (*OFF-state*).

Later, in [173,174,181,186] it was found that modeling the channel holding and idle times as exponentially distributed is not accurate. For channel holding times lognormal distributions and for inter arrival time Weibull or gamma distributions show better results. Furthermore, correlations between

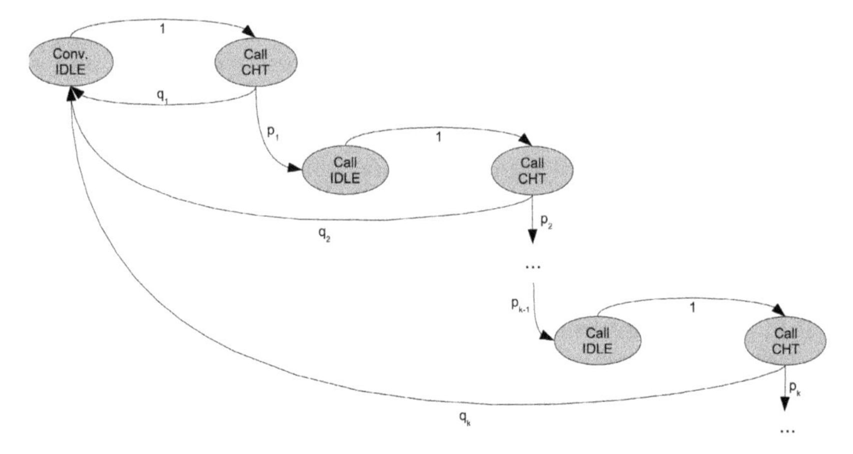

Figure 5.1 n-state (On-/Off-Model) (semi-) Markov model

calls with respect to short- and long-range dependencies have been examined. For the call holding times there were no correlations found whereas interarrival times showed dependencies (cf. [181]). Nevertheless, the studies still imply a two state model with non-exponential state holding time distributions.

While studies of the 1980s consider the traffic of a single channel (talk group), newer studies focus on traffic of the complete trunked radio systems (multiple talk groups mixed). However, the lack of examining single channels results in less accuracy when modeling the traffic of one talk group. A single talk groups has probably only a minor impact on infrastructure-based networks and their components. However, when infrastructure-less multi-hop networks are considered, the traffic of each talk group is of major interest. In [147] and [150] traffic models for single talk groups in disaster area and first responder scenarios are introduced. Details concerning these models are described in Section 5.4.1.

As the units in public safety scenarios do not move randomly over the whole simulation area, appropriate traffic distribution models are needed as well. There is one model [148] available for disaster area scenarios. This model will be described in Section 5.4.2.

5.4.1 Traffic Models

In [147] voice traces from a large catastrophe maneuver were analyzed. The time series were examined with respect to dependencies and heavy load peri-

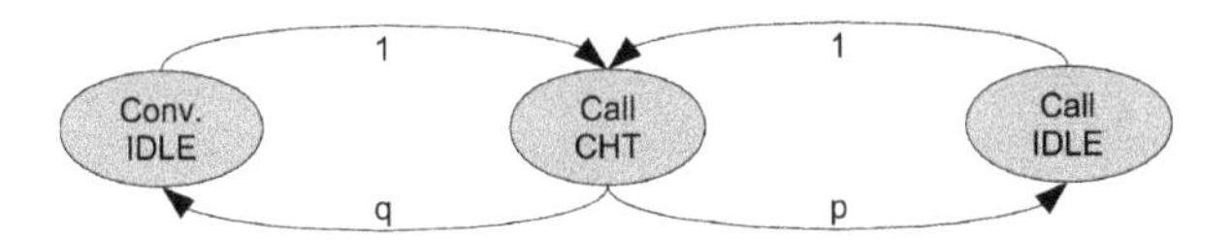

Figure 5.2 3-state (semi-)Markov model

ods. Based on the conversational dependencies of the call idle times, a n-state (semi-)Markov model was provided (see Figure 5.1). After a conversation idle time $Conv\_IDLE$ a new conversation starts with at least one call with channel holding time $Call\_CHT$. After this call there can be either a short idle time $Call\_IDLE$ (with probability p_1) or a longer one $Conv\_IDLE$ (with probability $q_1 = 1 - p_1$). In case of a long idle time we return to the first state ($Conv\_IDLE$). These two idle times distinguish between calls of the same conversation and different conversations. In case of a short idle time, the conversation has another call with channel holding time $Call\_CHT$. After this again there may be either a short idle time $Call\_IDLE$ (with probability p_2) or a longer one $Conv\_IDLE$ (with probability q_2) and so on (with probabilities p_k and q_k, respectively).

The transition probabilities p_k and q_k can be determined from the measured conversations calculating conditional probabilities as follows:

$$COL_k := \begin{cases} \emptyset & k \leq 0 \\ \{\text{Conversations : CallCount} \geq k\} & k \geq 1 \end{cases}$$

$$COE_k := \begin{cases} \emptyset & k \leq 0 \\ \{\text{Conversations : CallCount} = k\} & k \geq 1 \end{cases}$$

$$q_k := P\{\text{CallCount} = k \mid \text{CallCount} \geq k\} = \frac{|COE_k|}{|COL_k|}$$

$$p_k := 1 - q_k$$

COL_k are the conversations that contain more than k calls, while COE_k are the conversations that contain exactly k calls. The probability for a conversation containing at least k calls is the product of the probabilities $p_1, \ldots, p_{k-1}$, whereas exactly k calls is the product of the probabilities $p_1, \ldots, p_{k-1}, q_k$.

As the channel holding and call idle times are modeled identically for all calls, a simpler 3-state model (see Figure 5.2) can be used. After a call with call holding time $Call\_CHT$ (state in the middle) there may be either a short idle time or a longer one. Compared to the n-state model, the count of short

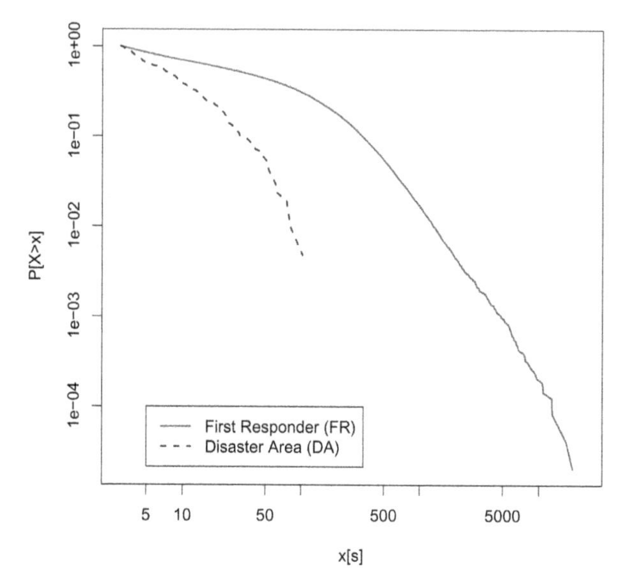

Figure 5.3 Comparison of Conversation Idle Time distributions of FR and DA time series via complementary cumulative distribution functions

calls in a row is not modeled explicitly due to a memoryless property of this model. The probabilities p_k and q_k are the same for different k. After k short idle times it is as probable that a long one follows as it is after the first short idle time. In theory there is no maximum number of calls per conversation in the 3-state model in contrast to the n-state one. However, the probability for a greater number of calls per conversation in the 3-state model is very small.

Different state holding time distributions were obtained for call holding times, call idle times, and conversation idle times by fitting the distributions to the traces. For the call holding times and conversation idle time the best fit was found in a lognormal distribution. For the call idle times the best fit was found in a the gamma distribution.

Furthermore, the load of a channel was varying and heavy load periods are determined. The heavy load periods are found to be mainly caused by variation at the conversation idle times. Intuitively it is quite obvious: heavier load implies more conversations which implies smaller conversation idle times. Thus, the variation of load and especially heavy load periods should be considered when fitting the distribution for the conversation idle times.

In [150] it was shown that traffic of first responder networks shows similar characteristics. Thus, the models are also applicable for modeling this traffic.

The main difference between disaster area and first responder networks was found to be the conversation idle times. See also the complementary cumulative distribution functions (CCDFs) in Figure 5.3 that visualize the probability distribution of the conversation idle times. We show CCDFs instead of cumulative distribution functions (CDFs) as we are more interested in the tail of the distribution. The load in catastrophe situations is higher than in first responder networks. The average traffic intensity of the first responder channel is smaller than the one of the disaster area channels.

Evaluations in [147, 149, 150] showed that n-state as well as three-state model generate realistic voice traffic for public safety scenarios.

5.4.2 Traffic Source Distribution Models

The modeling is based on an analysis of traces. This analysis showed that more than 92% of the conversations follow the star communication pattern (see [148] for further details). When modeling star communication, a larger number (around 50%) of the calls is originated by the head of the talk group. Thus, an equal distribution of traffic sources cannot be assumed. According to the star communication pattern all conversation are initiated or answered by the head of the talk group. Thus, roughly every second call can be assumed to be a call from the head. The peers that communicate with the head may be assumed to be uniformly distributed. According to the star communication pattern there is no reason why any group member other than the head should originate a larger number of calls.

To model the star communication pattern calls within one conversation are assigned alternatively to the head of the talk group and the (uniformly chosen) corresponding member. Whether the head or the group member starts the conversation is according to our measurements uniformly distributed. This approach does not model mesh communication patterns. More detailed models that also consider some conversations following the mesh communication pattern may be realized (cf. [148]). However, the impact of more detailed model on the results of simulative performance evaluation was not found to be significant.

5.5 Impact of Realistic Models

In this section, we show that using the detailed models for tactical scenarios has an impact on the results of the performance evaluation. For doing this, we provide results for two different kinds of evaluations. At first, we will examine the impact of the disaster area models on the simulative network performance analysis of routing protocols. Then, we will show the impact on the performance evaluation of different topology control strategies.

5.5.1 Impact on Routing Protocol Evaluation

For the simulations we consider a generic modeling of mobility, traffic, and distribution, as well as detailed models for disaster area modeling described above.

As one sample for a performance evaluation we show results for a routing protocol evaluation in MANETs. Simulations were performed using ns-2.29 [157]. We used the TwoRayGround propagation model with a uniform communication range of 100m. As MAC protocol, IEEE 802.11b Wireless LAN with 11 Mbit/s was chosen. As routing protocols we used the On-Demand Multicast Routing Protocol (ODMRP) [189], GRID [176], and Simple Flooding.

For the "generic" standard modeling, we chose Random Waypoint mobility model [172], constant bitrate traffic, and uniform node distribution. For the disaster area scenario, mobility, traffic, distribution of the nodes were modeled following the models described above. In analogy to a maneuver there are five local multicast groups and one global command channel. We modeled the voice traffic according to the Mixed-Excitation Linear Predictive enhanced (MELPe) [158] codec (developed for tactical communication) with 2.4 kbps. This results in 21 bytes IP payload every 67.5 ms. Parameters concerning velocities, pause times as well as bitrates were chosen equally for the generic modeling as well as the disaster area one.

To examine whether the disaster area models have an impact on the performance of the network, we measured the *timely packet loss rate*. The timely packet loss rate is a combination of packet loss rate and packet delay – for each talk group:

$$tPLR_G^{\delta} = \frac{\sum_{\forall n \in G}(P_{n,G} - R_{n,G}^{\delta})}{\sum_{\forall n \in G} P_{n,G}}$$

where $P_{n,G}$ is the number of application data packets that a node n as part of a talk group G should have received. $R_{n,G}^{\delta}$ is the number of application data packets of the talk group G received by nodes n with a delay smaller than δ. The traffic simulated is voice traffic. Thus, the packet delay δ has a decisive impact on application data. A packet that is too late will not be of any use for the voice data communication. We assume a packet with a transmission delay

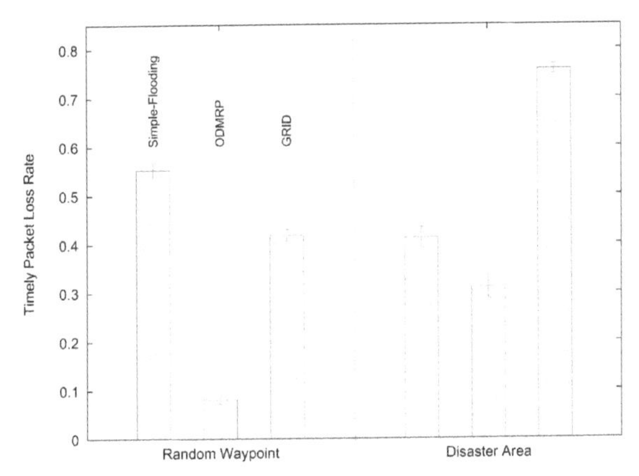

Figure 5.4 $tPLR_6^{82.5ms}$ for the different routing protocols and scenarios

larger than a threshold δ as lost. $tPLR^\delta$ can be analyzed for each node or for a whole talk group. Based on the MELPe codec we choose $\delta = 82.5$ ms.

Figure 5.4 shows the distribution of $tPLR_6^{82.5\text{ ms}}$ for the nodes of one talk group (group 6 – the global command talk group) using mean and 0.95 confidence intervals for the 20 replications. It is quite obvious that detailed modeling has a severe impact on the results of performance evaluation. The results for the generic modeling show large (>50%) loss rates for Simple Flooding. ODMRP as well as GRID show better results. Both try to overcome the well known broadcast storm problem of simple flooding (cf. [185]). GRID produces routing messages perpetually in a proactive manner. ODMRP shows lower packet loss rates as it only sends routing messages on demand. However, the disaster area modeling shows different results. Simple Flooding shows significant lower loss rates, while the loss rates for GRID and ODMRP raised by more than 20 and 30%, respectively. The different results are caused by the area-specific modeling of mobility, traffic, and traffic source distribution. Due to high node density inside the tactical areas packet losses occur. Simple Flooding makes it even worse by causing a lot of retransmissions. However, these retransmission have also an advantage: the losses caused by the high node density can be compensated. Using GRID worsens the overload due to its perpetual exchange of routing messages.

Overall, the realistic modeling shows a severe impact on the results of the simulative performance evaluations. From the results with generic modeling one would follow that ODMRP or GRID help to gain lower packet loss rates.

But the evaluations with a scenario specific modeling show different results: ODMRP is not that good and GRID causes even more losses than Simple Flooding. Thus, it cannot be assumed that all performance evaluation results obtained with generic models will hold for realistic public safety scenarios.

5.5.2 Impact on Topology Control Evaluation

During the past years, several topology control strategies as well as algorithms based on these strategies have been proposed (see [180]). The goal of these strategies is to adjust the transmission power of the nodes. By doing so, the topology of the network should be controlled in a way that there are no partitions and a maximal spatial reuse.

One strategy refers to the so-called *max-degree-topologies*. The basic idea is to adjust the transmission range in a way that the number of neighbors for each node lies in a range of $[k_{\min}; k_{\max}]$. If a node has more than $k_{\max}$ neighbors, it lowers its transmission power. If it has less than $k_{\min}$ neighbors, it raises its transmission power.

Another strategy relates to the so-called *nearest-neighbor-topologies*. They are explicitly designed to generate topologies where the k nearest neighbors are connected based on a nearest-neighbor-graphs. There is an edge between two nodes if one of the nodes belongs to the k nearest neighbors of the other. Due to this characteristic the number of neighbors of a node may be larger than k if the nodes are not distributed uniformly. The goal of this approach is to provide each node with at least k neighbors.

In order to examine the impact of these strategies, we use a disaster area scenario and compare it to the most commonly used Random Waypoint scenario. We examine topologies (node positions) at distinct times. Thus, we generate 20 movement traces of 3000 s length for the Random Waypoint and the Disaster Area scenarios. The topologies in these traces are examined every 50 s resulting in 1200 topologies for each mobility model. For each of the topologies we derive proximity graphs using the the topology control strategies described with different parametrization. As parametrization for max-degree and nearest-neighbor, we consider aiming at four, six, and eight neighbors. For reference purposes we consider fixed range topologies with 75 and 100 m fixed circular communication range.

In general, it is desirable to have a non-partitioned network. Only if the network is non-partitioned each node can communicate with any other at any

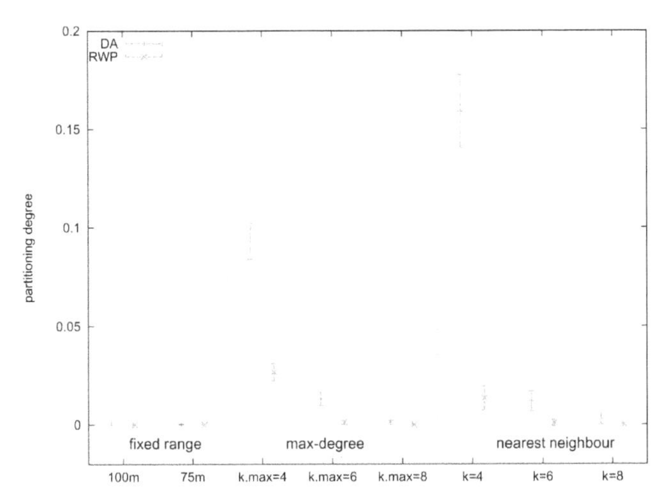

Figure 5.5 Partitioning degree (average and 0.95 confidence intervals) for different strategies and mobility models

point in time. The *partitioning degree* is defined as:

$$P = \frac{|\{(i, j) \in S^2 : i \neq j \wedge i, j \text{ not interconnected}\}|}{|S| \cdot (|S| - 1)}$$

where S is the set of all nodes and where *not interconnected* means there is no (multi-hop) path between two nodes. A large partitioning degree means that there are many nodes which are not connected and cannot communicate. Connectivity is really important, especially in public safety scenarios: each unit shall be able to communicate any time it is necessary. Thus, there is high demand for reliable, non-partitioned communication systems. Figure 5.5 shows the partitioning degrees (average and 0.95 confidence intervals over different topologies) for different topology control strategies and mobility models. The topologies without topology control that use fixed communication ranges show low partitioning degree. This could be expected, as we considered only quite large ranges where the dense networks are sufficiently connected. The max-degree and nearest-neighbor topologies show non sufficient partitioning degrees especially for small targeted numbers of neighbors. Choosing k or k_{max} as low as possible makes sense to get a higher spatial reuse. However, if the nodes are non-equally distributed, the number of neighbors nearby may be sufficient for the targeted number of neighbors. Thus, the network may be partitioned. The max-degree achieves lower partitioning degree compared to the nearest neighbors due to a specific property. Some nodes may excessively raise their transmission power, because they have not enough links. Their closest neighbors have already $k_{\max}$ links, and thus they cannot raise their transmission power. Due to the rigorous limit of $k_{\max}$ a node with not many other nodes nearby has to excessively raise its transmission power till he finds others with the same problem. This yield suboptimal topologies, especially if the nodes are non-equally distributed. There are some nodes (the outliers) that use a very high transmission power while others quite nearby use low transmission power. This is suboptimal for the overall power consumption as well as for the spatial reuse.

The DA model shows worse partitioning degrees for both strategies. The DA model considers hierarchical organization and tactical areas. Thus, the nodes are not equally distributed. The nodes are concentrated in the tactical areas. Especially for small k a node may find the desired number of neighbors in his own tactical area. Thus, the network may be partitioned. Overall, tactical areas that are typical for public safety scenarios are a specific challenge for topology control strategies. Generic models (like RWP) that do not model specific node distribution typical for public safety scenarios show too optimistic partitioning degrees.

To examine which strategies yields suboptimal topologies concerning power consumption and spatial reuse, the asymmetry may be considered. The asymmetry is a measure for the inhomogeneity of the transmission powers used by the different nodes. It is defined as:

$$A = \frac{|\{(i, j) \in S^2 : i \neq j \wedge r_i \geq d_{ij}\}|}{|\{(i, j) \in S^2 : i \neq j \wedge r_i \geq d_{ij} \wedge r_j \geq d_{ij}\}|}$$

where S is the set of all nodes, d_{ij} is the Euclidean distance between i and j, and r_i the transmission radius of i. Intuitively, the asymmetry is the fraction of uni-directional to bi-directional links. A value of one means that this node has no uni-directional links. The larger the value, the more uni-directional links a node has. Uni-directional links cannot be used by the nodes as e.g. routing protocols often assume bi-directional links. However, the uni-directional links may interfere transmissions of other nodes. When a node carrier senses the medium it may not realize that the medium is already used due to an uni-directional link. Thus, the larger the asymmetry, the more collisions occur, and the worse the capacity of the network.

Figure 5.6 shows mean and maximum asymmetry (average and 0.95 confidence intervals over different topologies) for the Random Waypoint and the Disaster Area scenarios with different topology control strategies.

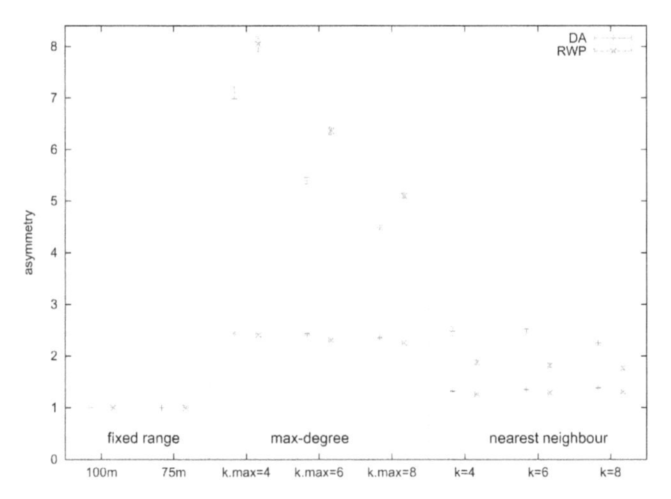

Figure 5.6 Mean and maximum asymmetry (average and 0.95 confidence intervals) for different strategies and mobility models

Fixed range topologies show asymmetry values of one, as all nodes have the equal transmission power. The max-degree strategy shows, compared to the nearest-neighbor strategy, much larger asymmetries. The reason for this is its rigorous limit of k_{max} that causes an excessive raising of transmission powers as described above. Random Waypoint shows larger maximal asymmetry due to the more spread out and randomized distribution. This figure verifies the interpretation given above. For the nearest neighbor topologies, the DA model shows, compared to RWP, little larger mean asymmetry and significant larger maximal asymmetry. The reason for this lies in the concentration of nodes in tactical areas. This may have bad impact on the capacity.

When modeling public safety scenarios, specific models should be used. Using generic models may yield idealized results e.g. concerning connectivity and asymmetry. Especially for topology control strategies, public safety scenarios that realize hierarchical organization and tactical areas are challenging.

5.6 Summary

In this chapter we presented an overview of models available for realistic modeling of public safety scenarios. Detailed models for mobility, traffic, and traffic source distribution have been developed. In two different sample evaluations the impact of the detailed modeling on the network performance evaluation was examined. The disaster area models allow a realistic scenario modeling. Results from generic, frequently used models show significantly different results. Thus, modeling with specific models for public safety scenarios discloses new information and has a significant impact on the results of the performance analysis.

There is a need for further models that complement the existing ones concerning the scenarios considered (e.g., models for large scale first responder scenarios do not exists). Beside this, a big challenge is the validation of existing and new models based on real measured traces. Furthermore, models for specific public safety scenarios tend to be quite complex. Thus, the usability of the models and with this enabling the community to use these models is another challenge.

6

Inter-Network Operability/Multi-System Resource Sharing Method for Disaster Relief

Hoang Nam Nguyen, Kazunori Okada and Osamu Takizawa

Disaster Management and Mitigation Group,
National Institute of Information and Communications Technology, Japan;
e-mail: nguyen@nict.go.jp

6.1 Introduction

Natural disasters, e.g. earthquakes and tsunami might occur at anytime and anywhere without accurate forecast. In disaster stricken areas, disaster recovery and relief activities have to be deployed fast and reliably which demand high efficient emergency communications capabilities. After disasters occur, emergency communications can be provided via different wireless communications networks e.g. satellite, cellular mobile and ad hoc networks in order to allow reliable communications to rescue teams, survivors/victims and other related people such as volunteers, press reporters, etc. [190]. Severe disasters, e.g. earthquakes might destroy buildings and infrastructure in a large disaster area. In a disaster stricken area, four zones can be identified including: disaster fields where search and rescue activities are carried out, evacuation places where survivors reside temporarily, the headquarter of disaster management and outside areas. These zones have different requirements on emergency communications and information. According to these requirements, emergency communications networks for disaster relief can be designed appropriately.

Figure 6.1 shows the classification of source, destination and content of emergency communications regarding four disaster related zones. For ex-

N. Marchetti (ed.), Telecommunications in Disaster Areas, 147–170.
© 2010 *River Publishers. All rights reserved.*

TX \ RX	Disaster field	Evacuation place	Headquarter	Outside areas
Disaster field	[P] Report situation, Relief request, Transfer command [U] Rescue teams etc. [C] Control signal, Audio, Video, Image	[P] Surveillance [U] Camera to victims [C] Video, Image	[P] Report situation, Relief request [U] Rescue teams etc. [C] Audio, Video, Image	[P] Reply to inquiry of safety, Press, Report disaster situation, Relief request [U] Victims to concerned persons Reporters to media Local headquarter to national headquarter [C] Text, Image, Audio, Video
Evacuation place	[P] Control surveillance [U] Victims to camera [C] Control signal	[P] Information for evacuation facilities [U] Victims, volunteers, responsible persons [C] Text, Video, Image	[P] Information of disaster situation. Request for facilities [U] Responsible persons to headquarter [C] Text, audio, Image	
Headquarter	[P] Transfer command, surveillance [U] Headquarter to rescue team [C] Control signal, audio, video	[P] Inquire of evacuation facilities [U] Headquarter to responsible persons [C] Text, Video, Image	[P] Information sharing [U] Persons in charge [C] Text, Video, Image	
Outside areas	[P] Inquiry of safety information, Press	[U] Concerned persons to victims Media to reporters	[C] Text, Image, Audio, Video	

P: Purpose U: User C: Content of communications

Figure 6.1 Sources, destinations and content of emergency information

ample, the headquarter needs to exchange command and relief requests with rescue teams. The interactions might be voice commands, surveillance images which cause high demand in network accessibility. Another example is that survivors located in disaster fields and evacuation places need to communicate with concerned people such as family members and friends, to inform them about their situation. This might cause a high required traffic load. In evacuation places, victims need to receive information about which evacuation activities are carried out and update other disaster-related information, e.g. about evacuation facilities can be used, volunteers who can ask for assistance. From the demand of emergency communications and information, an efficient communications infrastructure in disaster areas should be able to support different end terminals, e.g. fixed and cellular phones, PDA, computers and provide various communications modes such as mobile and ad hoc networking capabilities. The infrastructure should include heterogeneous networks, support various types of mobile terminals and be able to provide realtime and non-realtime services.

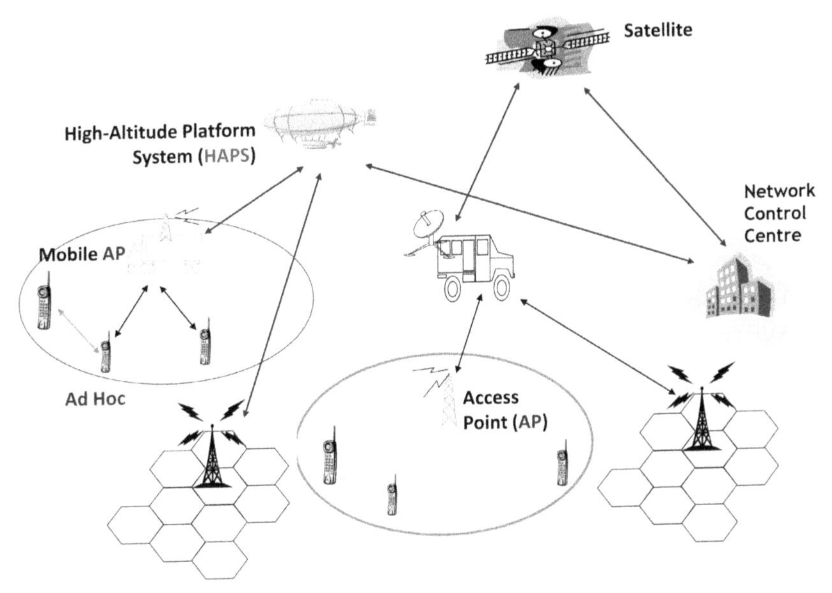

Figure 6.2 Heterogeneous emergency wireless communications networks

In a disaster area, major telecommunications facilities might be destroyed causing interruption of communications services. Due to the needs of immediate communications availabilities, emergency communications networks have to be set up in the shortest possible period. Comparing with wired communications, wireless communications is a much more efficient candidate due to the fast deployment of wireless transceiver stations, large coverage area and capabilities in supporting various types of end terminals. Nowadays, wireless terminals can be equipped with several wireless air interface radios, e.g. second generation (2G) and third generation (3G) cellular mobile communications, WiFi 802.11 and WiMAX as well as advanced networking technologies, e.g. ad hoc networking. Using multiple air interface radios and advanced wireless networking technologies, wireless emergency communications can increase the emergency communications survivability to victims.

Figure 6.2 shows an overview of wireless emergency communications deployment in disaster situations. Various wireless communications systems can be deployed in disaster areas, providing both space and terrestrial wireless communications to fixed and mobile wireless stations. We might assume that disaster areas are isolated with outside areas because backbone links were broken down. Satellite communications and High Altitude Platform System (HAPS) are used for connecting disaster areas with outside areas. From the rescue point of view, HAPS can be used for seeking victims who are not able to connect to other wireless networks because they might reside out of their coverage. In disaster areas, some places can be still covered by cellular mobile networks. In other places, cellular base stations (BS) could be collapsed thus access points (AP) using WiFi IEEE 802.11 technologies can be quickly set up to cover such places. In some locations, mobile APs are distributed to seek for victims. Victim's wireless terminals can access AP/BS directly or via ad hoc mode by hopping from terminal to terminal.

As radio resource is limited in terms of frequency spectrum and transmission power of wireless radio stations, an efficient emergency communications system should be able to provide capabilities of inter-network operation and radio resource sharing. In the system, a victim should be able to easily switch between different emergency communications networks. A victim shall also be able to access to any network having available radio resource.

Hybrid cellular and ad hoc communication model was proposed for emergency communications recently [190]. Integration of cellular–ad hoc communications systems is able to provide extension of cellular coverage and support self-survivability to mobile users. In such cellular-ad hoc systems, a mobile end terminal can operate in either cellular or ad hoc mode according to its situation. If the mobile terminal is not equipped with cellular radio, it can seek for connectivity to the emergency network by searching available neighbors via ad hoc interface. A mobile terminal connecting with a cellular base station can become a bridge of another mobile terminal to connect with the base station. Although ad hoc communications mode shows effectiveness in extending radio coverage and flexible connections capabilities, practical deployment of ad hoc mode is still facing social related challenges, e.g. a user might not be willing to spend its terminal power to help other users, security issues, etc. In this chapter, we will discuss details of a practical case in which we consider the radio resource sharing of cellular mobile networks in disaster situations.

Nowadays, mobile communication is popular in all countries, confirmed by the fact that the number of mobile subscribers has increased dramatically during recent years. Consequently to a disaster situation, a huge number of mobile subscribers might reside in disaster areas. That will result in a great demand of emergency mobile communications services. As discussed in [191], commercial wireless communications, e.g. cellular mobile networks still plays important roles in providing emergency communications in disaster areas beside other alternatives such as satellite communications and spontaneous wireless ad hoc networks. However, cellular mobile communications networks need to be equipped with enhanced functionalities in order to support emergency communications. Due to limited radio resource and capacity of wireless networks, particular resource management mechanisms are needed to achieve high system performance, e.g. low call blocking rate, in emergency situations. A dynamic call holding time mechanism has been proposed in [192] which is able to provide low call blocking rate in disaster areas. Another important issue is that in disaster areas, a mobile network might experience a number of base stations being destroyed, resulting in mobile subscribers in disaster areas that might not be served properly. With a huge amount of mobile subscribers, providing communications services to cellular subscribers becomes an important issue in disaster situations.

In the next section, we will present a survey of related research as well as practical systems or testbeds for emergency communications. After that, details of a multi-system access model and a radio resource sharing mechanism for 3G mobile networks will be presented. Section 6.4 will show the system performance results obtained by computer simulations. Conclusion remarks will be given in the last section.

6.2 Related Works

Recently there have been quite a lot of research efforts on emergency communications systems for disaster relief which focused on how to build a fast and reliable emergency communications system in disaster stricken areas. An emergency communications system consisting of heterogeneous wireless networks is considered as an efficient emergency communications infrastructure where end terminals used by victims and disaster responders are able to access more than one network and their communications with other parties is supported via inter-network operability. This section includes a brief survey of proposed theoretical researches and testbeds for inter-network operability and multi-system emergency communications. Here, we introduce illustrative theoretical researches which have dealt with issues of inter-network operability and multi-system resource sharing, which provide both spectrum and facility sharing, in various models such as a "cellular–public safety" system, a "satellite–mesh" system and a combined multiple wireless LAN system. We also introduce a testbed of "satellite–ad hoc" system and a proposal

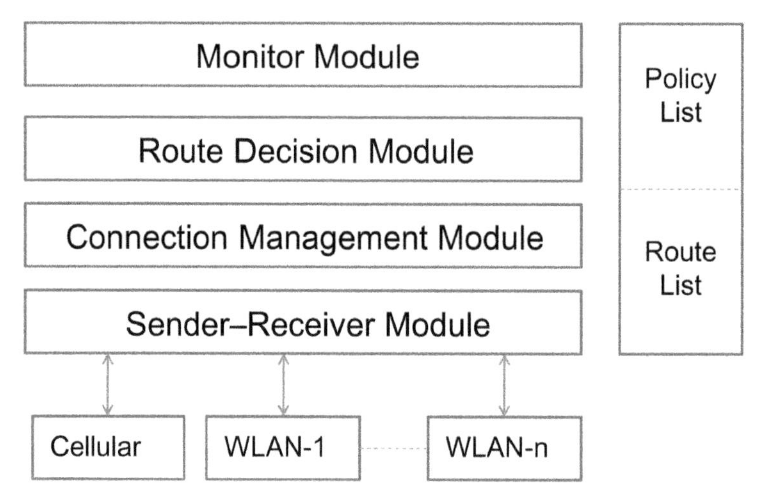

Figure 6.3 System architecture of combined wireless LANs for disaster relief

of Personal Digital Cellular (PDC)–Personal Handy-phone System (PHS) multi-system emergency access.

In [193], a disaster information system has been developed for a multiple wireless LAN based emergency system. A mobile node is equipped with multiple available wireless LAN and a cellular radios as shown in Figure 6.3. A communication path between nodes has multiple links and the most suitable link is selected based on different selection criteria such as the distance between nodes, transmitting power and operating frequency. In order to select the most suitable link and the best connections path, mobile nodes deploy a Monitor Module to monitor the changes of link quality, traffic load, distance. Policy List and Route List are used by the Route Decision Module for determining connections routes. As described in [193], connectivity between two nodes is supported by several links. When the current link throughput is decreased, the sender will choose a new link according to default configuration, throughput, power density, packet loss and the list of WLAN, in priority's decreasing order, i.e. the radio which has more available resources will be used with higher priority. An end-to-end path has one or more alternative routes which are used when the default route causes degraded throughput. In the research presented in [193], a prototype was built to evaluate the system functions and performance where several IEEE 802.11 WLAN and cellular CDMA 1x-EVDO were installed in a mobile node. When the distance

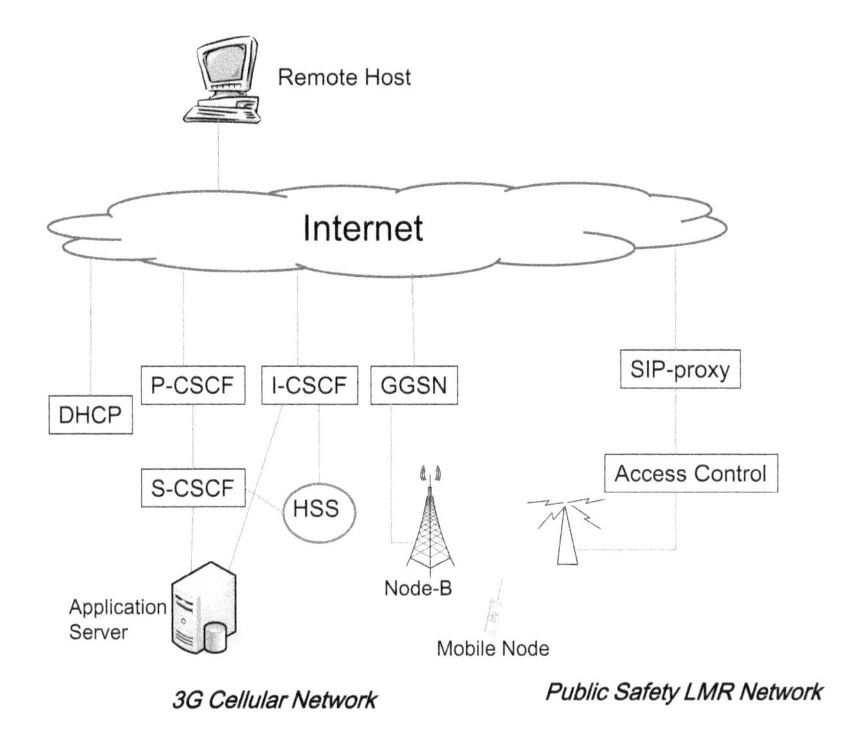

Figure 6.4 Inter-operability of wireless cellular and public safety LMR networks

between nodes increases, depending on the quality requirements of ongoing services, a node will select the most suitable link which provides guaranteed throughput. The prototype has validated the efficiency of this system and points out that dynamic routing and parameter values should be optimized in further research.

In another research found in [194], interoperability of public safety land mobile radio (LMR) with commercial wireless cellular networks for disaster response has been introduced. In this research, the interoperability shown in Figure 6.4 has been enhanced by exploiting Session Initial Protocol (SIP) and a joint radio resource management framework. Nowadays, LMR networks are used by public safety agencies for coordinating rescue teams and providing emergency response. Most current deployed LMR networks provide narrowband voice service and low-speed data services. While providing interoperability with cellular mobile networks, disaster responders can obtain a wide variety of benefits such as multimedia services and increased data rates which are not available in LMR systems. In the interoperable cellular and LMR networks, disaster responders can maintain their ongoing connections when they move out of the coverage of LMR networks by taking connection handoff to cellular networks, i.e. supporting the communications continuity. A novel SIP-based seamless handoff scheme was proposed for the system in order to reduce handoff delay. Considering Wideband Code Division Multiple Access (WCDMA) networks inter-operating with LMR networks, a joint radio resource management framework is designed aiming at keeping the handoff blocking probability below a target value. Based on this framework, authors of [194] have proposed an optimal radio resource management scheme that maximizes the radio resource utilization and guarantees service availability and continuity. In this radio management framework, services are distinguished in different classes. Incoming and handoff sessions are assumed to arrive according to a Poisson process while service duration is assumed to be following as exponential distribution. By using a semi-Markov decision process formulation and linear-programming based algorithm, the optimal resource management scheme can be computed.

Another research effort described in [195] is to design a mobile ad hoc satellite and wireless mesh networking for emergency communications scenarios. In this system, shown in Figure 6.5, ad hoc mesh networking plays an important role and mobile IPv6 is deployed for mobility management in order to support transparent and seamless movement of end terminals between local coverage areas. End terminals can be either public safety vehicles or private mobile terminals. Satellite communications is used for providing Internet connectivity to the disaster site. Vehicles which are equipped with Satellite Universal Mobile Telecommunications System (S-UMTS) devices become connecting points of ground ad hoc mesh networks with outside areas. Connectivity between vehicles is provided through mobile ad hoc mesh networking. A vehicle can act as mobile router in critical areas where mobile terminals form WiFi 802.11 and ad hoc networks. Mobile routers connect local WiFi 802.11 networks with the core mobile ad hoc mesh networks. IEEE 802.11s is assumed as the relevant standard for mesh and ad hoc networking in this architecture. The proposed hybrid system needs an efficient mobility management in order to provide seamless mobility to Public Safety

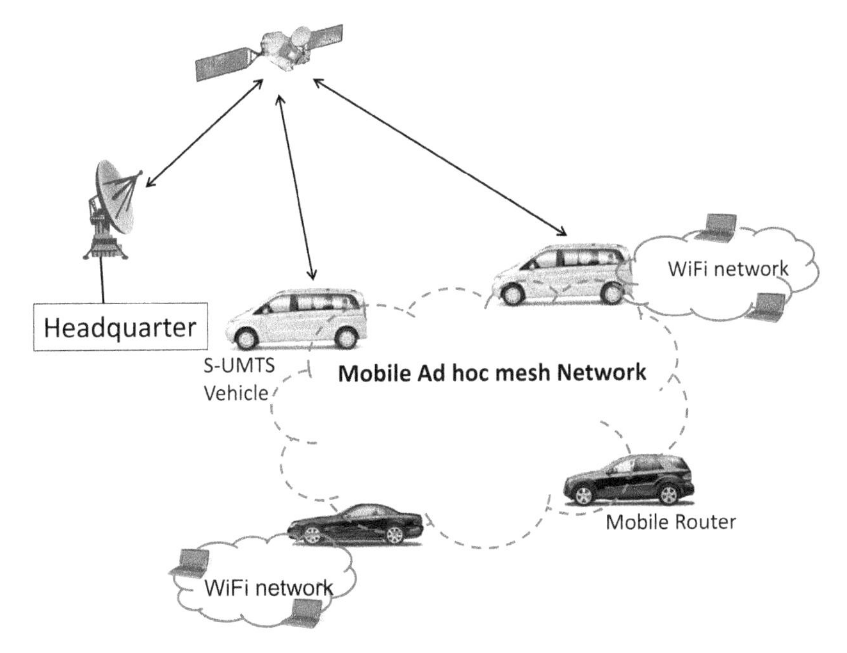

Figure 6.5 Hybrid satellite-mesh networking for public safety communications

units. In this research, proxy mobile IP version 6 (PMIPv6) has been exploited for this purpose. The delay of handoff has been analyzed according to mobility scenarios. Authors have shown that PMIPv6 can support efficient use of wireless resource and reduce handoff latency.

Beside theoretical research of multi-system access and interoperable networks for emergency communications, various efforts have been carried out in building test beds of emergency communications systems. As described in [196], an emergency system, denoted as DUMBONET and based on a combination of mobile ad hoc networks, a satellite IP network and terrestrial Internet has been implemented. The disaster stricken field might include several sites. Each rescuer uses a WiFi capable mobile device such as a laptop or a personal digital assistant (PDA). Rescuers residing in a disaster site connect with each other via ad hoc communications mode. One or several rescuers might be equipped with satellite communications supported devices to maintain the connectivity between disaster sites or between disaster sites and the command headquarter center. A virtual private network (VPN) is deployed in order to hide the network heterogeneity caused by different networking technologies including satellite, ad hoc communications and terrestrial Internet. DUMBONET also implements multimedia communications, sensor applications and human identification by face recognition.

Another testbed known as SKYMESH has been presented in [197]. SKYMESH is an ad hoc communication system which is as an urgent communication network backbone and built at 50–100 meters over the ground by using balloons. The advantages of this system include good line-of-sight, low interference and long transmission range. SKYMESH is used as an emergency communication network for large-scale natural disasters and for collecting information on disaster areas in order to support rescue, recovery and surveillance. Using balloons as transceiver stations is more appropriate than using helicopters or airships because maintaining balloons at fixed locations for many days is much easier at lower cost. Ad hoc communication is provided by using 802.11 WLAN and Optimized Link State Routing (OLSR) protocol. Experiment results show the efficiency of SKYMESH for emergency communications in large disaster areas.

In most research within emergency communications for disaster relief, self-configuration, self-recovery and fast system deployment are major crucial technical challenges. Ad hoc networking is considered as key technology for emergency communications systems in many researches, since it can be established quickly in disaster areas, when assuming that the telecommunication infrastructure has been destroyed. There are some research efforts on emergency communications provided by commercial mobile networks. Among them, network congestion in wireless emergency networks is another important research issue. As found in [198], a combined macro and micro cell overlay system for emergency communications has been proposed for emergency communications. The macro cell network is based on a PDC system whereas the micro cell network is based on PHS system. Authors have studied the performance of this overlay system in case of disaster under the assumptions of very high incoming traffic load. Mobile terminals are equipped with two radio interfaces (PDC/PHS) and emergency calls can be sent to any network. Performance results obtained by computer simulations showed that the high gain in terms of low call blocking probability can be achieved when providing multi-system access opportunity to mobile PDC/PHS terminals.

Continuing recent research of emergency multi-system access, a WCDMA-based emergency multi system access has been proposed [199]. In the next sections, multi-system radio resource sharing between 3G mobile cellular networks is presented including three major contributions. First we propose an emergency communication system for providing communication survivability to WCDMA mobile cellular subscribers in disaster situations. In the proposed emergency communication system, several WCDMA mobile networks are co-operated to provide emergency calls to survivors/victims in disaster areas. The reason to select WCDMA mobile networks for the system design is that in Japan, among more than one hundred millions mobile subscribers, three WCDMA mobile networks provided by NTT Docomo, Softbank and EMOBILE companies have about seventy four millions subscribers [200]. The second major contribution we propose is to guarantee communication quality by exploiting a mechanism of selecting the proper base station. While deploying this mechanism, a mobile subscriber always initiates emergency calls by sending call requests to the most appropriate base station having the strongest pilot signals which can belong to any network. The third major contribution is an extensive system performance evaluation by doing practical WCDMA simulations. The system model and admission control are described in the next section and performance results follow in Section 6.4.

6.3 Resource Sharing between 3G Mobile Networks

6.3.1 System Model

We propose a WCDMA-based multi-system access emergency system in which several WCDMA cellular networks are involved for the purpose of delivering emergency services to victims and survivors locating in disaster areas. Figure 6.6 illustrates how the system works; we give an example of an emergency communications system involving 3 WCDMA networks, denoted as Network-1, Network-2 and Network-3. These networks might be able to cover fully or partially of the disaster area depending on the number of active base stations (BS) after the disaster. The size of the overlapped area depends on the location of base stations and the cell geographical map of these networks. Each network has a different number of subscribers located in the disaster field, i.e. different incoming traffic load, thus different radio resource utilization. For example, Network-1 might be overloaded due to very high traffic density whereas Network-2 still has available radio resource for new coming traffic.

Consider a mobile subscriber (MS) which is a contract subscriber of Network-2. At its location in the disaster area, assume the MS is able to receive radio signals from base stations of all three networks (Network-1, Network-2 and Network-3). Now we consider what happens when the MS

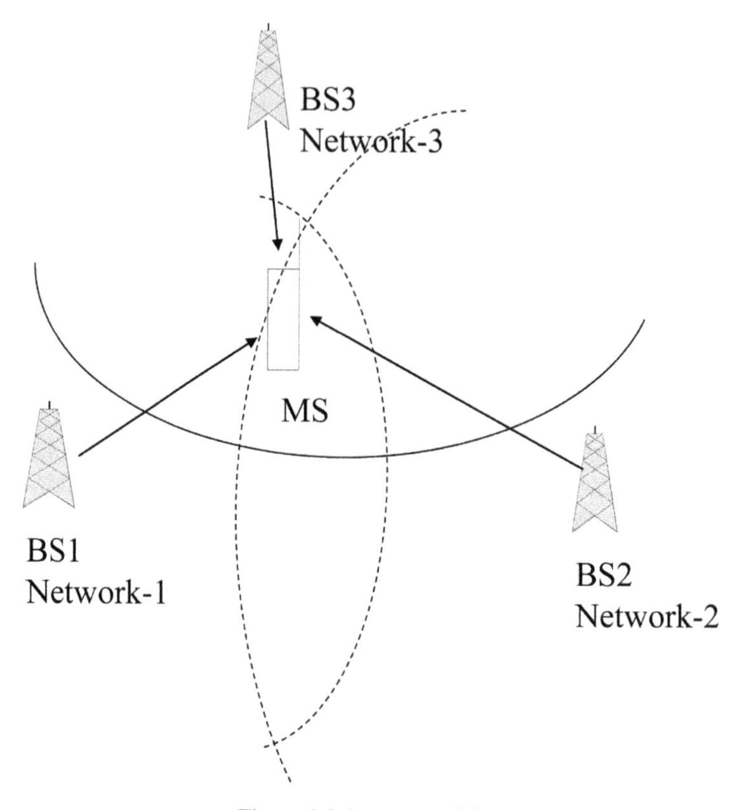

Figure 6.6 System model

needs to make an emergency call in a conventional emergency system. First, the MS will send a call request to its contract network, i.e. the Network-2. In the case Network-2 is overloaded, Network-2 will have to reject the call request due to lack of radio resource, otherwise the call request will be accepted. In this conventional system, a network which has low users density in the disaster area, i.e. low incoming traffic is able to provide calls at a high acceptance rate. However, a network which has high user density will have to reject much incoming call requests resulting in high call blocking rate. Network roaming can be a solution for this problem in which the MS will select another network to connect with permanently. However, the selected network might be also overloaded too thus roaming might not be useful. Roaming might cause additional signaling load whereas it does not guarantee that the subscriber's call demand would be accepted. Another technical problem is that in the conventional system, a MS might maintain its connection with serving BS of the contract network at low signal quality due to the long distance between the MS and the BS, whereas the MS can receive better signal strength from another BS of other networks. For instance, the MS in Figure 6.6 is closer to BS3 than to its contract BS2, i.e. the MS experiences a better quality for the signal received from BS3 than that received from BS2. Thus, when using the conventional system, the signal quality might not be well guaranteed resulting in low quality of service.

In order to fulfil disadvantages of the conventional emergency systems, our proposed emergency system provides MS' access capability to all available networks without roaming requirements. The principle is that in disaster areas, service providers will accepts call requests sent from any subscriber regardless of its contract network. In the proposed system, a MS is considered as a home user of its contract mobile network and a guest user of other existing mobile networks. BSs covering disaster areas will process call requests of guest MSs, i.e. call requests from disaster areas will have the highest priority. Instead of sending a call request to the contract network (Network-2) which is currently overloaded, the MS might send the call request to other networks (Network-1 and Network-3). Assume Network-1 has available radio resource, the call request will then be accepted by Network-1 and the MS will be able to make the necessary emergency call. Regarding the signal quality issue, the proposed emergency communication system utilizes a mechanism of selecting proper base stations in which a mobile subscriber will always try to connect to the base station which has the strongest pilot signal. More details on the access admission procedure and on the mechanism of selecting the proper base station are given next.

6.3.2 Emergency Access Admission Procedure

Figure 6.7 illustrates an emergency access admission procedure and the mechanism of selecting the proper base station. Consider a mobile subscriber K who is a contract (home) user of Network-2 and guest user of Network-1 and Network-3. The MS receives pilot signals from BS1, BS2 and BS3 which belong to Network-1, Network-2 and Network-3, respectively denoted as P_L^1, P_L^2 and P_L^3. Assume $P_L^3 > P_L^1 > P_L^2$.

When the subscriber K needs to make an emergency call, it will select Network-3 as the initial access network because the pilot signal of its BS3 is the strongest one. The MS sends a Call Request to BS3. BS3 verifies if the number of ongoing calls is less than both uplink and downlink capacity. If the

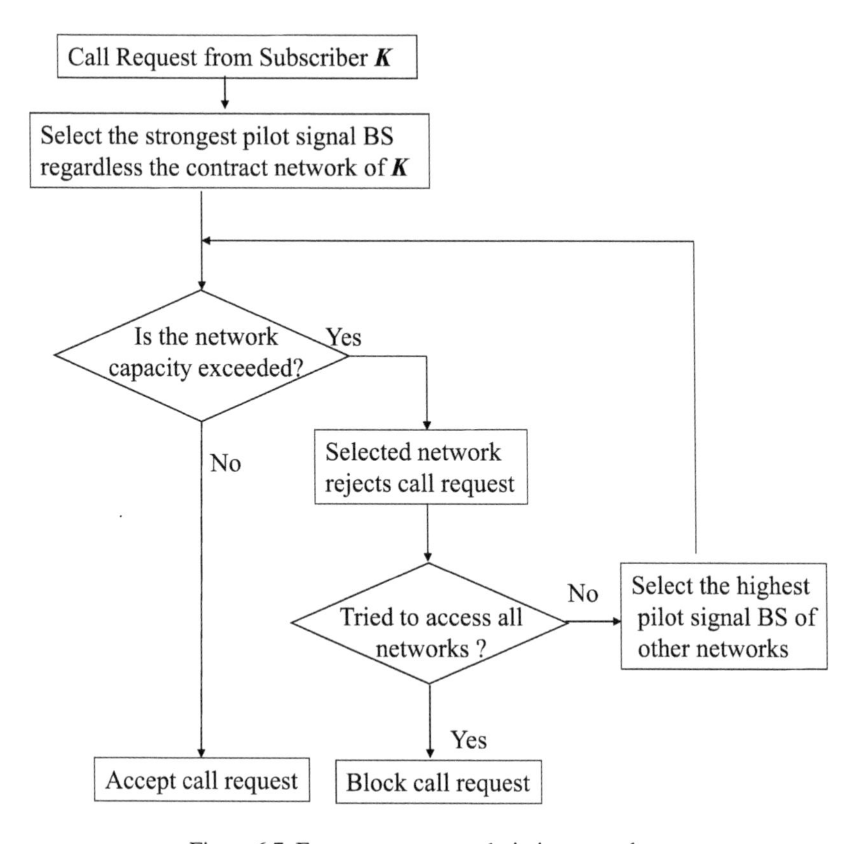

Figure 6.7 Emergency access admission procedure

system capacity is not exceeded, the call request is accepted by Network-3. In the case either downlink or uplink capacity of BS3 are exceeded, Network-3 blocks the call request and informs the MS about this decision. If the MS has tried to access other networks already, the MS will block the call request. Otherwise, the MS will select BS1 as the pilot signal of BS1 is higher than BS2. Then the call request is sent to BS1 and it will be accepted if the BS1 has available resource for this new call request. If not, the call request is sent to BS2 for the last attempt. If BS2 is also overloaded, the call request will be blocked.

The proposed emergency system brings the following advantages. First, the network with low incoming traffic will well support other networks with high incoming traffic, i.e. we have a better radio resource sharing. The second advantage is that because mobile subscribers always attempt to get connections to the BS which has the strongest signal, it brings high efficiency in user power consumption and induces low interference power to others in both uplink and downlink. Therefore, our proposed system is also able to support low system interference, i.e. able to provide high system capacity as WCDMA has an interference-limited capacity. Within the study of system performance of WDCMA-based emergency multi-system access, other important research issues are still open for further research. Among technical issues, signalling is one of the most important challenges. Signalling should be minimized so that mobile subscribers can switch between networks quickly and do not increase the signalling load too much. Another research issue is about the management policy; the mobile network operators should agree to join the emergency communication system when disasters occur. Under this policy, mobile device manufacturers need to add emergency related functionalities to mobile terminals. For example, mobile terminals should also be equipped with Emergency Mode so that in case of disaster, they are able to perform automatically the proposed mechanism of selecting the proper base station.

6.4 System Performance Evaluation

6.4.1 Simulation Model

We have developed a simulator which conforms technical specifications of WCDMA for UMTS systems given in [201,202] to evaluate the performance of the proposed emergency communications system. For simplification, the 19 hexagon cell model is used in the simulator. The system configuration includes three WCDMA networks with (X, Y) coordinates of cell centers that are (0, 0), (500 m, 500 m) and (−500 m, −500 m) for Network-1, Network-2 and Network-3, respectively, as shown in Figure 6.8.

In order to observe the benefits gained by radio resource sharing, the three following simulation scenarios have been implemented when studying system performance. Because there is not a detailed analysis of incoming traffic load of mobile cellular networks in case of disasters, system traffic load is studied under assumed low, medium and high levels of incoming traffic load as shown below.

- *"Symmetric parameter" scenario*: The WCDMA networks have identical cell size and traffic loads. Base stations are equipped with sector antennas. The cell radius is 1000 m. In this simulation scenario, we will compare performance of proposed and conventional systems,

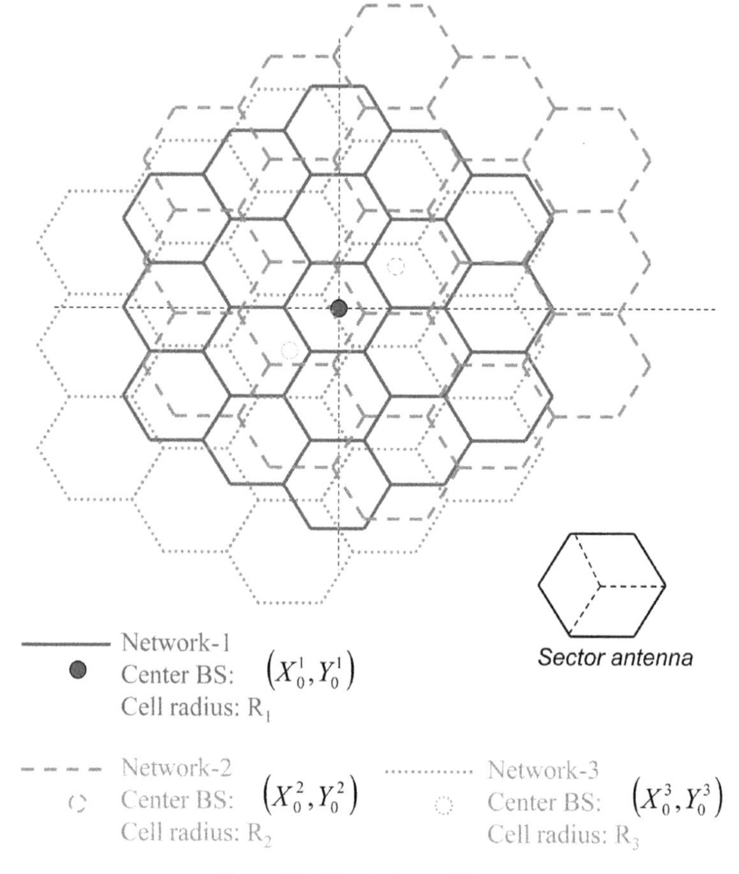

Figure 6.8 Simulation cell layout

where base stations are equipped with sector antennas. In order to show the effectiveness of deploying sector antennas and validate the simulation results, the performance of the proposed system deploying omni-antennas is also presented.

- *"Different cell size" scenario*: The WCDMA networks have different cell radius and identical traffic load. The cell radius is 1000, 1200 and 1500 m for Network-1, Network-2 and Network-3, respectively. Base stations are equipped with sector antennas. We will compare performance of individual WCDMA networks in the proposed and conventional systems.

Table 6.1 WCDMA simulation parameters

Parameter	Value	Unit
Carrier frequency	2000	MHz
BS's maximum transmission power	43	dBm
MS's maximum transmission power	21	dBm
BS's max power per channel	30	dBm/1Ch
MS's max power per channel	24	dBm/1Ch
BS's min power per channel	10	dBm/1Ch
MS's min power per channel	–56	dBm/1Ch
BS's initial power per channel	10	dBm/1Ch
MS's initial power per channel	–10	dBm/1Ch
BS antenna height	15	m
Target SIR downlink	8.9	dB
Target SIR uplink	7.1	dB
Required SIR downlink	6.4	dB
Required SIR uplink	5.5	dB
TPC rate	30	-/frame
Frame duration	20	ms
Duty factor on downlink	0.5	
Duty factor on uplink	0.5	
Bit rate	12.2	Kbps
Chip rate	3.84	Mbps
Interference margin	10	dB
f value	0.66	

- *"Different network load" scenario*: The WCDMA networks have identical cell size and different traffic loads. Base stations are equipped with sector antennas. The cell radius is 1000 m. Assume that Network-1 has a fixed traffic load of 1200 Erlang (low traffic load) and Network-3 has a fixed traffic load of 2000 Erlang (i.e. medium traffic load). The network load of Network-2 is variable. The performance of individual networks in the proposed and conventional systems is presented.

In all simulation scenarios, we assume victims/survivors equipped with WCDMA handsets locate randomly in each cell. Call requests occur according to a Poisson process. After a call request is accepted, the call duration is assumed exponentially distributed with the mean value of 120 s. Transmission power control is applied for both uplink and downlink at the standardized frequency of 1500 Hz. Other simulation parameters of the WCDMA model are given in Table 6.1.

The propagation model is ARIB vehicular model [202] as shown below:

$$L = 40(1 - 4 * 10^{-3}\Delta h_b)\log_{10} R - 18\log_{10}\Delta h_b + 21\log_{10} F + 80 \quad (6.1)$$

Table 6.2 SIR value at traffic load of 2800 Erlang

	Proposed			**Conventional**		
	Net1	Net2	Net3	Net1	Net2	Net3
UL-SIR	7.48	7.58	7.50	7.68	7.69	7.62
DL-SIR	9.91	9.94	9.99	10.54	10.53	10.54

where R is the distance between MS and BS, F is the WCDMA carrier frequency in MHz and Δh_b is BS's antenna height.

Call blocking rate and *access blocking rate* are considered in estimating the system performance and are defined as follows:

- *Call blocking rate* of a network's subscribers is defined as the ratio of the number of blocked call requests, which are not accepted by all networks, to the number of call requests from the subscribers.
- *Access blocking rate* of a network is defined as the ratio of the number of rejected access requests to the number of access requests sent to the network.

6.4.2 Performance Results

Figures 6.9 and 6.10 show the average call blocking rate and average access blocking rate, respectively, in the "Symmetric parameter" scenario. Traffic load of each network is varied from 1200 Erlang (low traffic load) to 2800 Erlang (very high traffic load). In general, the proposed system provides much better call blocking rate than that of the conventional system as shown in Figure 6.9. When the traffic load of each network is less than 2000 Erlang, the gain of call blocking rate is more than 30%. In the case of the very high traffic load, e.g. at 2400 and 2800 Erlang, the performance gain can still be more than 15%.

In contrast, the access blocking rate of the proposed system is higher than that of the conventional system, as shown in Figure 6.10. The reason is that in the proposed system, a WCDMA network receives a huge number access requests initiating from not only its contract (home) users but also guest users. Therefore when the system load is high, more call requests will be rejected in individual networks.

Performance results also show that deploying sector antennas provides much lower call blocking rate and access blocking rate than deploying omni-directional antennas. That is because in 3-sector cells, interference is much lower than in omni-antenna cells resulting in higher system capacity and higher SIR.

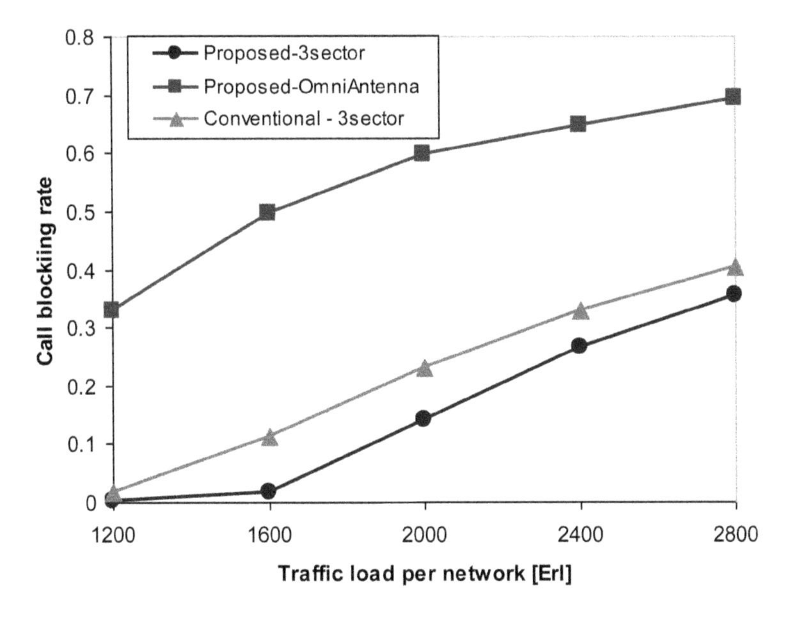

Figure 6.9 Call blocking rate in "Symmetric parameter" scenario

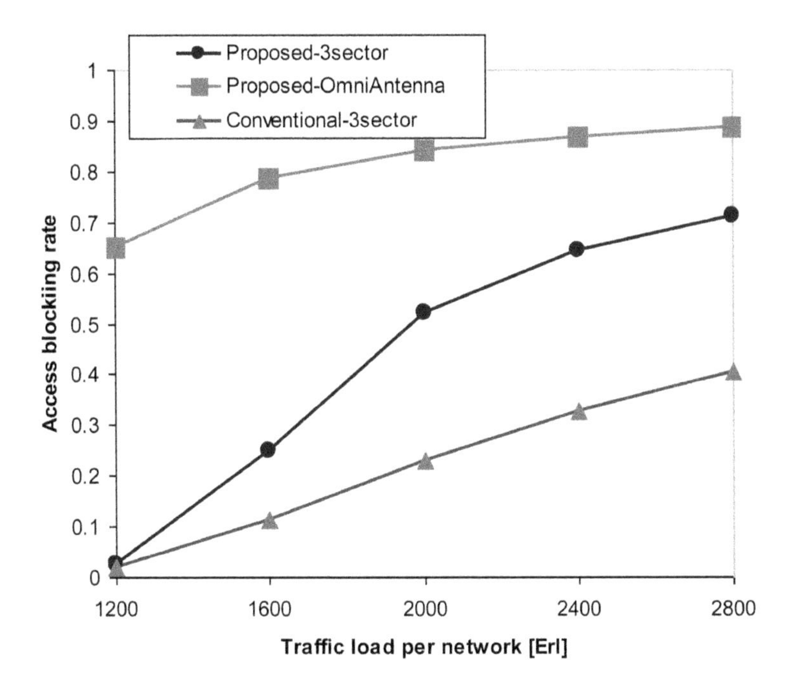

Figure 6.10 Access blocking rate in "Symmetric parameter" scenario

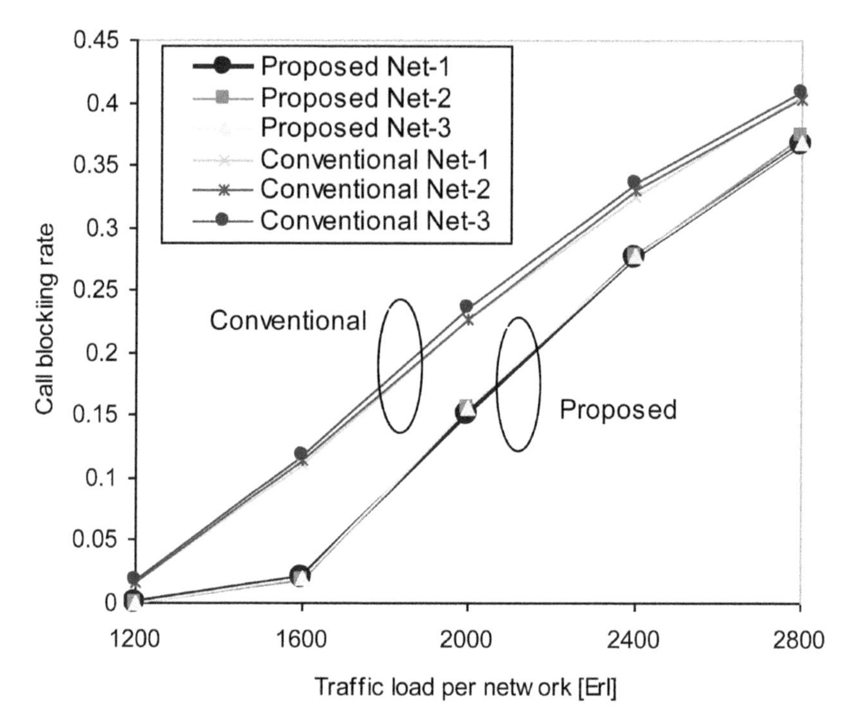

Figure 6.11 Call blocking rate in "Different cell size" scenario

Performance results of the "Different cell size" scenario are shown in Figures 6.11 and 6.12. Again, the proposed system outperforms the conventional system in terms of call blocking rate while it provides higher access blocking rate, due to similar reasons to those discussed previously. In both systems, the call blocking rate of individual networks is not so different although their cell size is not identical. That is because the difference of cell radius in larger (Network-3) and small (Network-1) cells is not much causing less degradation to SIR, as an example is shown in Table 6.2.

In terms of access blocking rate, individual WCDMA networks of the conventional system show similar access blocking rate as they have identical network load. In the proposed system, when the system load increases, the network, which has a larger cell size, has also a higher access blocking rate. In this case, Network-3 has the highest access call blocking rate. The reason is that the BS of Network-3 has larger coverage than that of Network-1 and Network-2. Thus it covers more cells of Network-1 and Network 2. In this

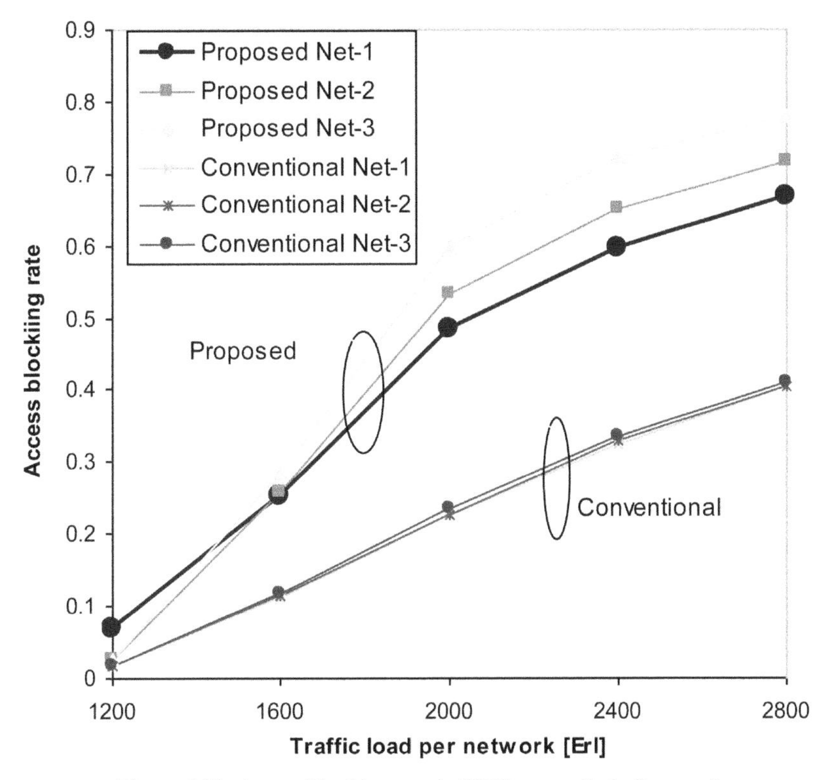

Figure 6.12 Access blocking rate in "Different cell size" scenario

case, BS of Network-3 will receive more call requests from subscribers of other two networks thus resulting in higher access blocking rate.

The aim of the "Different network load" scenario is to show how efficiently radio resource is shared in the proposed system. As shown in Figures 6.13 and 6.14, the proposed system provides similar call blocking rate and access blocking rate for the different individual networks. That means radio resource of WCDMA networks is shared with each other efficiently. In this scenario, Network-1 has low traffic load and shares available resource to Network-2 and Network-3. When the traffic load of Network-2 is more than 2000 Erlang, it can take radio resource of both Network-1 and Network-3 too. Therefore the call blocking rate and the access blocking rate of these networks are identical.

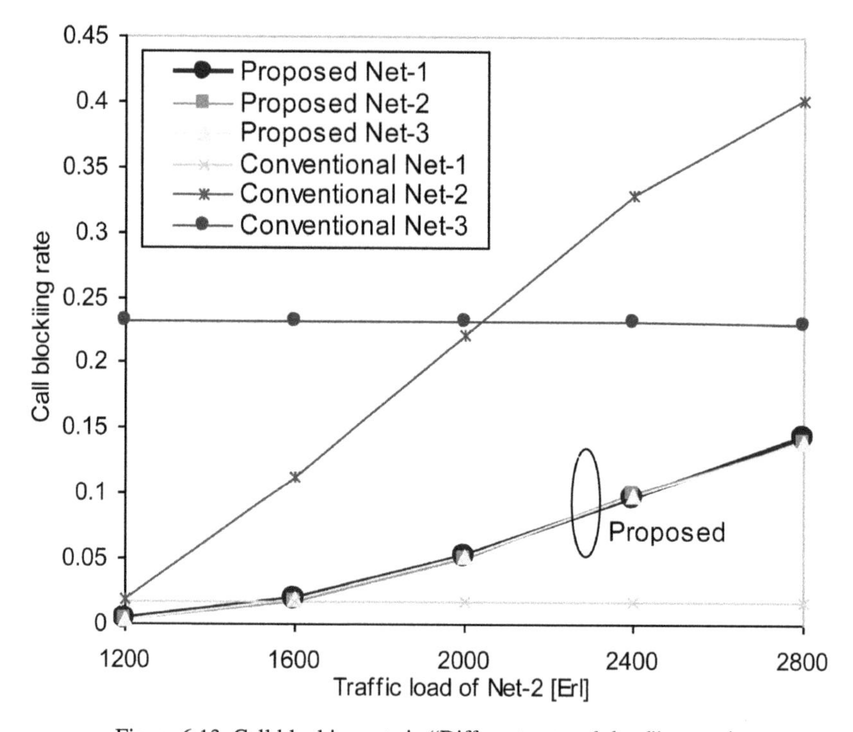

Figure 6.13 Call blocking rate in "Different network load" scenario

In contrast, in the conventional system, each network exhibits different performance. Network-1 and Network-3 have fixed traffic load thus their call blocking rate and access blocking rate are not varied. Available radio resource of Network-1 is not shared with other networks so that Network-2 shows worse performance when its traffic load increases.

6.5 Conclusion

In this chapter, we have discussed issues related to internetwork operability and multi-system radio resource sharing for supporting disaster relief and emergency communication. Many research efforts have been carried out for investigating heterogeneous emergency systems where the inter-operation of different wireless communications networks has been proposed for efficient emergency communication. Commercial cellular mobile networks, infrastructure-less ad hoc networks and satellite networks play an important

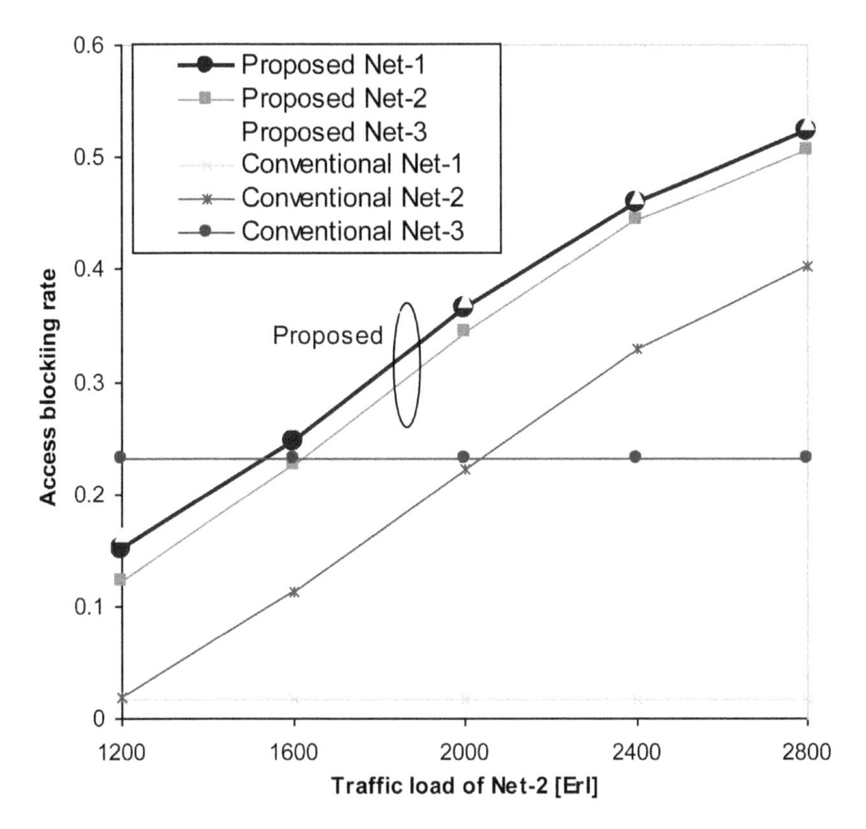

Figure 6.14 Access blocking rate in "Different network load" scenario

role in disaster mitigation and relief. In an emergency communication system, ad hoc mobile networks can be used for establishing self-deployment and auto-configurable emergency communication clusters in which survivors and disaster responders can communicate with each other in the hop-by-hop model. Communication between such clusters can be provided by either satellites or by airships. Further, if mobile terminals are equipped with several radios including WLAN and mobile cellular radios, mobile users can get better communications quality and higher survivability. Commercial cellular mobile networks become important facilities for providing emergency information services to disaster survivors, as the number of mobile subscribers is very high in most countries. System performance benefits gained by radio resource sharing between 3G mobile networks have also been presented, showing the need of future cooperation between mobile operators in case of disasters. Network facilities and mobile terminals should also be equipped with emergency supported functionalities. For further research, system performance evaluation using practical incoming traffic information needs to be carried out. The emergency scenario, in which base stations of 3G mobile networks are destroyed, is also an interesting research topic. Design of mobile terminals and system test beds is needed to improve and validate the proposal of multi-system resource sharing method presented here.

7

Location-Aware Cognitive Communication Systems for Public Safety

Ali Gorcin[1], Khalid A. Qaraqe[2], Hasari Celebi[2] and Huseyin Arslan[1]

[1]*University of South Florida, Tampa, FL, USA;*
e-mail: agorcin@mail.usf.edu; arslan@eng.usf.edu
[2]*Texas A&M University at Qatar, Doha, Qatar;*
e-mail: {khalid.qaraqe, hasari.celebi}@qatar.tamu.edu

7.1 Introduction

The introduction of cellular systems and evolution of analog wireless communications systems to digital systems opened an era of hundreds of millions of wireless communications users around the globe. Cellular phones are used in the first place in case of extreme situations or emergency cases and rescue-safety operations can benefit from prevalence of cellular wireless technologies during and just after extreme situations like hurricanes, earthquakes and collapse of buildings or mines. However, events around the world such as unfortunate attacks of 9/11, hurricane Katrina, Indonesian tsunami, and devastating Pakistan floods revealed problems about legacy wireless communications systems from a different aspect: Wireless communications of all kinds cease to work in extreme situations because of the damaged infrastructure, congestion caused by overwhelming request for communicating and abnormal communication parameters, e.g. drastic changes in air interface because of whether conditions during natural disasters. Establishing communications under severe conditions of extreme situations to detect and locate survivor communications or to maintain reliable communications between the first responders of various public safety organizations require reconfigurable and intelligent communications techniques. For in-

N. Marchetti (ed.), Telecommunications in Disaster Areas, 171–211.
© 2010 *River Publishers. All rights reserved.*

stance, the radio should sense the operational environment of communication and autonomously adjust its parameters accordingly to execute these tasks.

Another important issue is interoperability of public safety organizations or first responders. Wireless communications infrastructure may not satisfy the requirements of large operations that are needed during the spread natural disasters and events because, the number of the first responders increases with increasing population but, allocated resources and technology does not change by the time. Hence, it becomes nearly impossible to communicate even for the first responders of the same organization because of insufficient resources such as frequency spectrum. Besides, if there is no common infrastructure or technology defined to cover all public safety organizations in one region, it becomes nearly impossible to coordinate first responders which finally leads to inefficient resource usage and loss of crucial time in the very first response phases of the events.

Cognitive radio is an emerging wireless technology which can introduce methods for efficient, secure, and reliable communications between different public safety units during extreme situations. Adaptability and sensing features of cognitive radios present solutions to the problems mentioned, e.g., opportunistic spectrum usage and interoperability. Moreover, location awareness method can be explored to estimate the victims and first responders locations. The signals from the devices of the victims who are trying to reach out can be detected; their signal types can be identified by using spectrum sensing methods. Furthermore, first responders' wireless communications devices can communicate with the victims' devices even though original core telecommunications network may be down by benefiting from the adaptiveness feature of cognitive radio systems or the location of the victims can be estimated extracting required information from the received signals.

7.2 Background and Current State of the Art

One of the best cases related to the complexity of the issues of public safety communications is the US public safety communications system. There are more than 74,000 public safety organizations using various wireless communications technologies and frequency bands in the US according to National Public Safety Telecommunications Council (NPSTC). The interoperability issues of these organizations is a well known problem: first, effective communications among the first responders, i.e., police and fire departments during the emergency; second, cooperative communications that can enable these public safety organizations from different disciplines to make jurisdictions to work together in emergency preparedness and response. Even though the frequency spectrum allocation and licensing policy of Federal Communications Commission (FCC) since 1930s [203] seems to be the source of problems in the first place, the vertiginous evolution of wireless communications technologies should be also indicated as another reason for this divided map of organizations. There are some efforts to solve these issues which can be categorized in two paths: (i) regulatory issues and (ii) technological challenges.

In 1994, FCC and National Telecommunications and Information Administration (NTIA) established Public Safety Wireless Advisory Committee (PSWAC) to evaluate the wireless communications needs of local, tribal, State, and Federal public safety agencies through the year 2010, identify problems, and recommend possible solutions [204]. In 1996, PSWAC concluded it's study emphasizing three important problems of public safety communications systems: (a) lack of resources – primarily spectrum – because of increasing number of users, (b) interoperability problem due to incompatible radio equipment and use of multiple frequency bands, (c) requirement for benefiting from newly emerging wireless technologies to achieve more efficient communications between first responders and to take advantage of high data rate applications. Later in 1997, NPSTC was formed to encourage and facilitate implementation of the findings and recommendations of the PSWAC. However, ineffective rescue operations and unorganized relief efforts during hurricane Katrina revealed that immediate action is required to establish interoperability and cooperative communications between the first responders. After a series of steps taken by regulatory organizations, FCC is currently managing the operation on dedicating a fixed communication frequency band for public safety organizations in the US [205].

On the other hand, wireless communications technologies evolved in a fast manner through the end of twentieth century; introduction of cellular technologies and local area networks boosted the evolution from analog to digital. Rapidly increasing number of users and conversion of the voice oriented applications to multimedia applications boosted the requirement for efficient, flexible and adaptive communications systems. Hence, during the 2000s, wireless systems evolution followed the path from software radios through the software defined radios to cognitive radios which have sensing, learning and adaptation features [206].

Detecting and locating victims in a disaster area through their wireless devices and the nature of the mentioned problems above require research for intelligent, secure and robust communication techniques. The core of possible solutions relies on research for wireless channel modeling of extreme cases and also on secure physical layer reconfigurability [207, 208]. Reliable and solid solutions can be brought to these problems by conducting advanced research and by developing reliable applications depending on cognitive radio (CR) technology.

Traditional communication system design is based on allocating fixed amounts of resources to the user. Adaptive design methodologies, on the other hand, typically identify the requirements of the user, and then allocate just enough resources, thus enabling more efficient utilization of system resources and consequently increasing capacity. Even though there have been studies and definitions on radio systems which are computationally intelligent to choose and support multiple variations of wireless communications systems [209, 210], a widely accepted terminology is introduced by Joseph Mitola III through software defined radios (SDR) [211] and later on cognitive radios. The term Cognitive Radio is defined by Joseph Mitola III in his PhD dissertation entitled: "Cognitive Radio: An Integrated Agent Architecture for Software Defined Radio" [212] as:

> The term cognitive radio identifies the point in which wireless personal digital assistants (PDA) and the related networks are sufficiently computationally intelligent about radio resources and related computer-to-computer communications to:
>
> 1. Detect user communications needs as a function of use context, and
> 2. Provide radio resources and wireless services most appropriate to those needs.

Cognitive radio has been perceived as an extension of software defined radios earlier but later in [213], FCC suggested that Cognitive Radio also should include enhancements on spectrum awareness:

> A cognitive radio (CR) is a radio that can change its transmitter parameters based on interaction with the environment in which is operates. The majority of cognitive radios will probably be software defined radios (SDR), but neither having software nor being field programmable are requirements of a cognitive radio.

Even though there is no consensus on the formal definition of cognitive radio, the concept has evolved to include various meanings in several contexts. One of the studies that proposes a new wide area network (WAN) called unlicensed WAN which deploys cognitive radios that declare their coarse communication requirements before the network is established and source allocation is handled by the communication node or base station [214] of the WAN. Another main aspect is related to autonomously exploiting locally unused spectrum to provide new paths to the spectrum access. Cognitive radio technology is widely studied related to spectrum access management issues by introducing secondary users in licensed bands. [215] discusses in detail how opportunistic spectrum usage can be achieved through cognitive radios and a collaborative spectrum sensing methodology is proposed depending on the sensing of the frequency spectrum with multiple devices in coordination.

Some of the communications systems such as the ones in the cellular bands and ISM bands have dynamic usage patterns which do not use the allocated frequency spectrum continuously in time and space as static users like TV emitters and radio wave transmitters. This fact may lead to opportunistic spectrum usage by communication devices with spectrum sensing capability. Deployment of cognitive radios as secondary users in these bands is also another aspect of achieving spectral efficiency [216]. Moreover, currently some companies are working on implementation of cognitive radios that will work on TV broadcasting bands. There are also some challenges and limits investigated at the physical layer level of the communications stack. These challenges can be listed as geographical limits, issues related to signal detection, noise and quantization uncertainty [217]. [218] listed three key aspects of Cognitive Radio discussing some physical layer limitations and applications of wide band cognitive radio systems:

- Sensing – A cognitive radio must be able to identify the unused spectrum segments,
- Flexible – A cognitive radio must be able to change signal frequency and spectrum shape to fit into the unused spectrum segments,
- Non-interfering – A cognitive radio must not cause harmful interference to the primary users.

The advantages of CR technology over legacy systems stems from the introduction of flexible, reconfigurable and re-programmable wireless communications systems. One of the major impacts of these systems is seen on interoperability issues on both military communication systems and public safety communications. However, there are some challenges from the aspect of public safety applications; especially on the security issues and convincing the regulatory bodies to allocate specific frequency bands for the realization of these wireless communications technologies for public safety communications [219, 220]. In the light of these developments, National Public Safety Telecommunications Council (NPSTC) which is a federation of organizations whose mission is to improve public safety communications and interoperability, joined SDR Forum which is the supporting organization for SDR and CRs to come up with a solution to the interoperability problems of public safety agencies. NPSTC Software Defined Radio Working Group published the "Cognitive and Software Radio: A Public Safety Regulatory Perspective" report [221] in June 2004 employing cognitive radio as solution addressing FCC to solve the problems regarding allocation of frequency bands, testing and security. Unfortunately, one year later, the interoperability problems of public safety agencies became a major issue during the event of hurricane Katrina. Immediate action was taken by FCC by submitting a report to U.S. Congress addressing the interoperability problems [222]: as the first action, 700 MHz frequency band (108 MHz of spectrum from 698–806 MHz) is allocated to public safety and Congress directed that TV broadcasters should complete transition to digital broadcast technologies and vacate the spectrum in order to accommodate wireless commercial and public safety uses of the spectrum until February 17, 2009 [205]. For the long term, the FCC [222] also indicated smart and cognitive radios as a first level commercial technology for the solution of interoperability issues of public safety communications systems that will operate in these new bands. Beside these activities, Homeland Security initiated SAFECOM; a communications program providing research, development, testing and evaluation, guidance, tools, and templates on interoperability of communications related issues to local, tribal, state, and federal emergency response agencies. SAFECOM published a report [223] detailing public safety communication device functional requirements which can be covered by cognitive radio technology. Moreover, SDR Forum also constituted public safety special interest group and published [224] recommending SDR and cognitive radio technologies for SAFECOM requirement satisfaction and advanced spectrum allocation at 700 MHz frequency band public safety operations. Beside these activities, Jesuale and Eydt [203] discussed implementation of policy based cognitive radios for public safety and industrial land mobile radio (LMR) bands currently in use. Introduction of cognitive radios with the cooperative sensing capability for public safety communications by applying spectrum polling techniques which accumulate available spectrum information is proposed. However, it is also indicated that an extensive research and development process is required for cognitive radio technologies asserting this process will be more beneficial, more realistic and much cheaper than dislocating all current public safety communication agencies into a single band. Jones et al. [225] detailed results of broad spectrum measurements made in public safety band in Howard County, Maryland, USA. The objective was to investigate the feasibility of public safety bands from the aspect of opportunistic spectrum access. Energy detection approach depending on spectrum sensing methodology of cognitive radio technology is employed as signal detection technique. The output of the measurement results is discussed taking the impacts of three characterizing parameters into consideration: (i) energy threshold which is used to detect the wireless signals, (ii) spectrum sensing period that is the time passing between the initialization of each sensing operation, and (iii) keep-off time which is defined as the minimum time duration over which no primary activity is detected before the secondary users access a channel. The measurements conducted proved useful in examining issues such as spectrum sensing for dynamic access, primary user signal detection, adjacent channel interference detection, receiver sensitivity, and policy performance with local and cooperative sensing.

It is also important to mention about the European and collaborative efforts for public safety communications interoperability since last two decades. Terrestrial Trunked Radio (TETRA) is a digital trunked mobile radio standard developed by the European Telecommunications Standards Institute (ETSI) [226]. The purpose of the TETRA standard is to meet the needs of traditional Professional Mobile Radio (PMR) user organizations such as public safety, transportation, government, commercial, military, etc. Deployed first in 1997 (Release I), TETRA uses time division multiple access (TDMA) and full duplex voice communications. Most importantly, it employs two different air interfaces, between the base station and radio terminals (1) and the Direct Mode Operation (DMO) interface (2) allows terminals to operate in local radio nets without support of main framework. Release II brought improvements on voice coders/decoders and also improved data rates up to 538 kbps for 150 kHz bandwidth and 64-QAM modulation. Several companies and research institutions also initiated EUROPCOM project which aims to investigate and demonstrate the use of ultra wide band (UWB) radio technology in emergency situations to allow the precise location of personnel to be displayed in a control vehicle and simultaneously improve communications reliability [227]. It is planned as an extension to previously deployed TETRA system to achieve traceability for first responders in extreme cases that can lead to loss of their lives. Furthermore, it is aimed to develop methodologies to detect and trace the victims in the operation area by benefiting from penetration of UWB signals through obstacles such as walls. European

Telecommunications Standards Institute (ETSI) and The Telecommunications Industry Association (TIA) agreed to co-operate for the constitution of mobile broadband specifications for public safety. The means for this activity have been provided in the form of a partnership project called MESA, originally known as the Public Safety Partnership Project (PSPP), which constitutes the legal and operational framework ensuring a swift progress of results. The aim of the project MESA is to produce globally applicable technical specifications for digital mobile broadband technology, focusing initially at the sectors of public safety and disaster response [228].

Some agencies in the US, such as National Institute of Justice (NIJ), also started to support research institutions, universities and companies to develop cognitive radio systems for public safety systems [229]. One of these studies proposes a set of algorithms called cognitive engine to bring in intelligence to an SDR based hardware platform [220]. It was intended to demonstrate that it is possible to achieve spectral interoperability through cognitive radio technology. Another study proposes an interim system design solution which selects the communication standard available, i.e., TETRA, TIA-603 in an adaptive manner. This approach propose solutions to the interoperability problems by introducing two paths, one depending on current signal processing technologies and approaches like software communication architecture (SCA) of Joint Tactical Radio systems (JTRS) program and another depending to open source technologies like micro Clinux [230]. It is also indicated that in the future, developed architectures should be implemented fully on cognitive and software defined radio.

Even though cognitive radio technology found one of its initial application as spectrum agile radio for public safety communications, diverse inherent features of cognitive radios introduce challenging research and implementation opportunities for public safety communications. System level requirements of public safety cognitive radio are detailed from the aspects of achieving awareness, learning and decision making in [231]. Jesuale [232] introduced employment of cognitive radio technologies depending on the primary (current public security agencies) and secondary (later joined commercial entities) user approach in current public safety frequency bands. Interoperability and efficient usage of the frequency bands are expected to be achieved by the deployment of new digital cognitive radio technologies both for primary and secondary users. Hoeksema et al. [233] and Pawelczak et al. [234] propose an adaptive ad-hoc cognitive radio emergency/disaster relief network. First the service and system requirements are defined on data messages, real-time voice, still picture, real-time video, and remote control. The proposed cognitive radio system is a spectrum agile radio diverted from IEEE 802.11 standard based on spectrum polling and distributed channel assignment (DCA) concepts. Media access control layer (MAC) design and algorithm implementation are also detailed. A reconfigurable hardware architecture for emergency networks, depending on orthogonal frequency-division multiplexing (OFDM) based cognitive radio is introduced in [235]. This architecture is intended for wireless nodes which are expected to employ adaptive bit loading and power loading for spectrum sensing purposes. Hardware implementation of such cognitive radio architecture in a flexible and energy efficient way is studied benefiting from the heterogeneous and reconfigurable technologies of field programmable gate arrays (FPGAs), digital signal processors (DSPs), and application specific integrated circuits (ASICs). Besides, Green and Taylor [236] has introduced a real-time cognitive radio test bed for interoperability and spectrum sensing experiments. A multiple-input multiple-output (MIMO) public safety cognitive radio hardware platform is proposed to achieve interoperability. In the context of public safety, it is shown that cyclostationarity detection methodology for user parameter (e.g. modulation type, center frequency) identification can be used for spectrum sensing purposes. On the other hand, Wang et al. [237] has proposed a wireless emergency communications framework depending on two-hop relaying and cognitive radio technology. According to the proposed approach, in case of an emergency situation, if the wireless infrastructure fails to provide the necessary service to its users (mobile terminals in this case) coverage expansion is achieved by relaying and frequency lowering achieved by cognitive radio technology. Finally, a cognitive radio model for self-identification of emergency situation for mobile handsets is proposed in [238, 239]. A software based switching mechanism between an ad-hoc mode called emergency mode and normal mode is introduced. This approach aims to enable disaster survivors' communication with other available nodes in case of absence of the centralized network. The proposed solution is based on the cognitive radio technology using game theory and artificial neural networks. Such an approach for self-identification of emergency situation for mobile handsets requires major changes in the current commercial handsets and the infrastructure.

7.3 Cognitive Radio for Public Safety: Opportunities and Challenges

FCC ruled to the evacuation of 700 MHz frequency band under a partnership of public safety organizations and private sector for public safety communications as an urgent action after the unfortunate event of Hurricane Katrina [240]:

> Under the Public Safety/Private Partnership, the Public Safety Broadband Licensee will have priority access to the commercial spectrum in times of emergency, and the commercial licensee will have preemptible, *secondary access* to the public safety broadband spectrum. Providing for shared infrastructure will help achieve significant cost efficiencies while maximizing public safety's access to interoperable broadband spectrum.

The FCC's rule making for public safety communications proposes gathering of all public safety communications operations into a single band. Major consequences of this rule making can be categorized into five folds:

(i) Interoperability of public safety agencies are expected to be achieved by gathering all services into a single frequency band. The expenses of nationwide exchange of current analog and digital legacy public safety radios with new digital technologies can be compensated by opening some portions of this band to commercial use. The commercial use of the spectrum as the consequence of public safety/private partnership approach is expected to cover huge expenses of the technology transformation. Even though it may take time to convert all the devices of public safety organizations to the devices which operate in a single band, this approach avoids the deployment of interim policy based cognitive radios which are proposed as the short term solutions to the interoperability problems [230].

(ii) Gathering all public safety organizations in a single band does not directly satisfy the requirements of effective communications among the first responders and does not provide the cooperative communications that will enable the public safety organizations from different disciplines to collaborate in emergency preparedness and response. Introduced wireless communications technologies should be able to come up with solutions to these problems satisfying requirements of SAFECOM.

(iii) Secondary access to the spectrum requires sensing, awareness and decision making for the devices that will be deployed in the public safety/private partnership frequency band.
(iv) Coverage area of a wireless communications system is a function of operating frequency band. Legacy systems operating on VHF and UHF bands easily cover the wide geographical areas of rural places with sparse population. Even though there seems to be no problem in urban places with dense population, deployment of systems working on 700 MHz frequency band can be problematic for rural areas because the coverage area at the 700 MHz band will be narrower than VHF and UHF bands [203].
(v) FCC's static frequency spectrum allocation and licensing policy divides the spectrum into chunks and assigns each chunk to a e specific service. This approach will lead to congestion in the future as happened in the past. Introduced technologies should be open to apply spectrum polling and dynamic spectrum usage to overcome these problems.

When the evolution of the wireless communications technologies is taken into consideration, cognitive radio is the major technology to satisfy the requirements of public safety communications. Cognitive radio technology with features like adaptiveness, sensing, and intelligence can introduce a wide variety of application options for public safety beyond proposing solutions to the problems such as interoperability. Therefore it would be appropriate to consider the possibilities and opportunities that the employment of cognitive radios brings to public safety communications from a wider window. It should be investigated that how public security communications can benefit from cognitive radio technology, instead of considering only satisfying the public safety requirements and bringing solutions to the problems. Identification of the communication medium (wireless channel), physical layer reconfigurability to overcome communication congestion and to satisfy the interoperability requirements, security considerations, and location awareness are the main issues that should be considered in the context of public safety cognitive radio systems.

7.3.1 Wireless Channel Models for Emergency Situations and Extreme Cases

The transmission medium, so called channel in which information carrying signals propagate through is the most important phenomenon that should be taken into consideration while designing communications systems. It can be clearly said that without accurate channel information it is not possible to design a reliable communications systems for a certain medium. The channel for wireless communications systems in the most basic sense is air or atmosphere. However, in reality, wireless signals should penetrate, be reflected and pass through obstacles such as buildings, hills and mountains, lakes, etc. Hence, wireless communication channels exhibit different characteristics depending on the specific propagation environment. For instance, the propagation and fading characteristics of urban and rural areas are different from each other. Through the evolution of wireless communications, realistic wireless channels for urban, semi-urban and rural areas are modeled either by conducting extensive measurements or by developing simulation models. Therefore wireless communications channel characteristics should be studied in the first place. This approach would give more insight into the implementation of qualitative interoperability and the cease of communication phenomenon in extreme situations.

Research and development of wireless channel models for extreme cases such as forest fires, hurricanes, earthquakes, mines collapses, etc., can be conducted by simulating these communication environments. The parameters that should be taken into consideration are considered and weighted corresponding to variations called large scale fading, small scale fading and a medium-scale effect called shadowing [241]. Therefore, certain parameters should be defined and the modeling of the channel should be conducted adjusting these parameters accordingly. For example, while the number of the buildings and the density of the structuring are the main parameters for channel models of environments such as rural areas, the modeling parameters for extreme cases can be more case dependent. Even though the environmental parameters can also be considered, the importance should be given to parameters such as change in clutter, rain density, wind speed, heat, density of dust, humidity, number of the wireless communications users' equipment, current wireless infrastructure, types of deployed wireless and wire line communications systems, change in the location of base station or communication node antennas and so on. Some of the effecting parameters indicated as hexagonal shapes and their direct or indirect effects to the wireless communications channel modeling parameters are linked in Figure 7.1.

The outcomes of extreme case wireless channel modeling research should guide physical layer reconfigurability research on functional block design, algorithm and parameter selection level of adaptiveness. Location awareness research to estimate the locations of the victims can also benefit from the

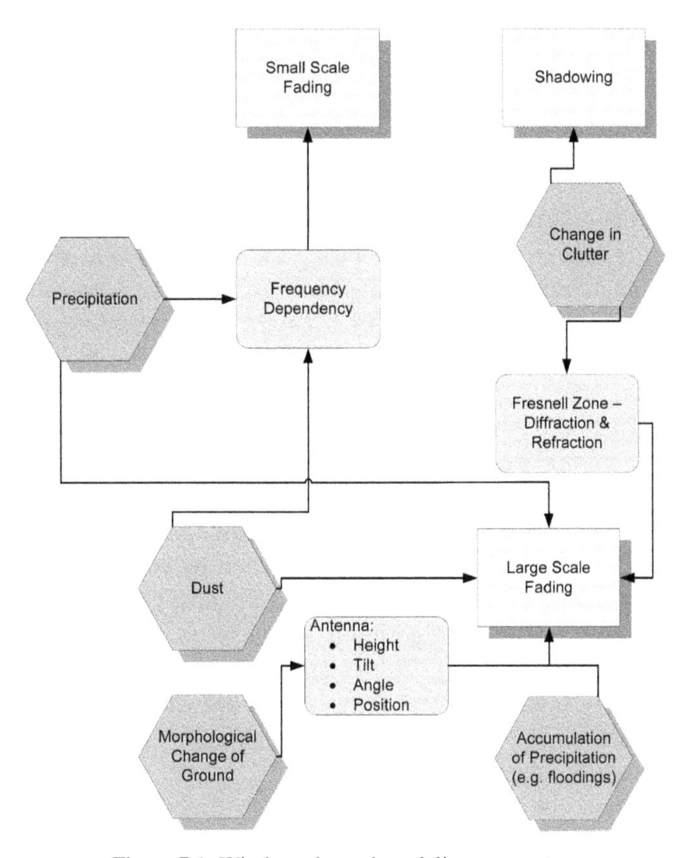

Figure 7.1 Wireless channel modeling parameters

outputs of modeling research by becoming able to isolate the channel effects from the incoming signal.

7.3.2 Physical Layer Reconfigurability

Communication congestion originating from the nature of the extreme situations and interoperability problems stemming from resource insufficiency of communications systems are the two main factors that cause the cessation or inefficiency of public safety communications systems during extreme situations. The solution to these problems can be found by understanding the wireless channel (which is detailed in the previous section) and by developing the physical layer reconfigurability.

One of the widely studied features of the cognitive radios is their ability to sense the communications environment, especially the spectral occupancy. Cognitive radios are affiliated with the public safety communications from the aspect of spectrum sensing and opportunistic spectrum usage, in general. However, cognitive radios can also autonomously observe and learn about the radio environment, generate plans, and even correct mistakes. These features give the cognitive radio technology the flexibility and adaptability to overcome the problems of public safety communications systems.

As the radio senses the operational environmental features such as frequency band occupancy, wireless channel characteristics like fading, time and frequency selectivity, delay and Doppler spread, subsequently, it can be aimed to achieve physical layer reconfigurability through autonomous selection of appropriate algorithms for functional blocks of frequency, signal power level and waveform selection, modulation-demodulation methods, encoders-decoders, interleaving and de-interleaving, etc. Depending on the requirements of the operational environment and by comparing the environmental features with the developed channel models, cognitive radio should be able to select appropriate algorithms for each functional block to execute the required tasks successfully. This is called algorithm selection level adaptability [242]. For instance, interleaving and de-interleaving functional block can change interleaving method depending on the fading values or depending on the selectivity of the wireless channel.

Besides that, for each algorithm of each functional block, there are some parameters that can be adjusted. For example, coding rates for coders-encoders can be tuned depending on the requirements. This is also called algorithm parameter level adaptability. Another aspect of physical layer reconfigurability research is the introduction of an ad-hoc networking mode for public safety radios. Ad-hoc networking can be achieved through the implementation of a predefined common communication channel such as the un-addressed voice channels in TETRA or broadcast voice channels for talk groups of traditional public safety wireless communications systems [233–235]. Another research area includes investigation of cooperation methodologies which can be based on an ad-hoc communication mode or original core communications network. Cooperation between public safety cognitive radios can be achieved taking advantage of physical layer reconfigurability. If the communication network is available on the field, the CRs of the different first response organization can tune into the communications parameters such as frequency and modulation type dictated by the network and establish cooperative communications. On the other hand, the ad-hoc

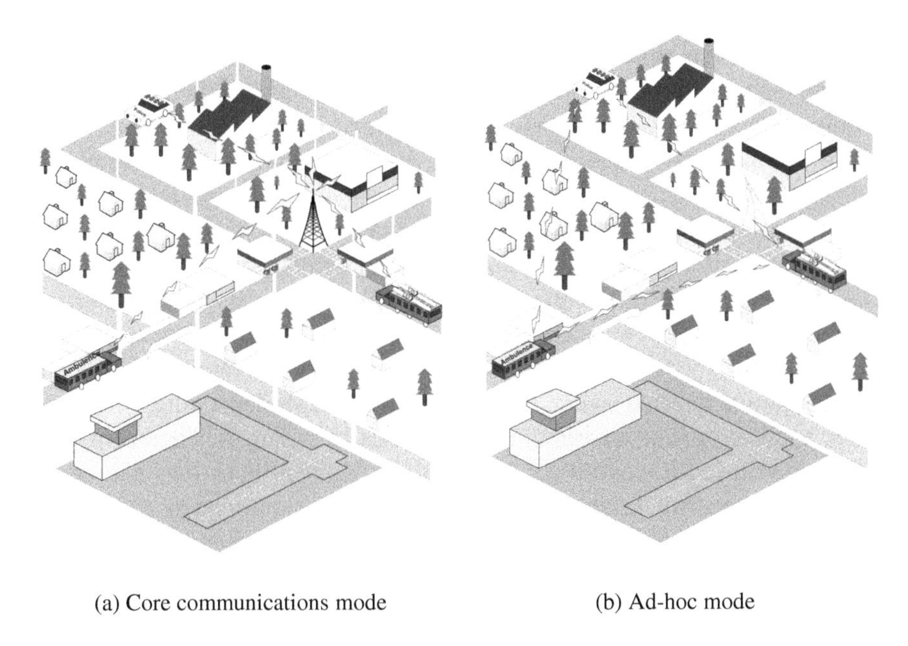

(a) Core communications mode (b) Ad-hoc mode

Figure 7.2 Adaptation of communication modes

networking mode can be established when there is no network support. In this case, cooperative spectrum sensing can be possible with contribution of multiple public safety wireless communications equipments and as the radios in the environment detects each others presence, ad-hoc communications can be established. Moreover, the crucially important issue for victims and first responders is that the communications between first responders will not be ceased if the core communications network is down. The ad-hoc mode and core communications modes are depicted in Figure 7.2.

7.3.3 Security Considerations

The voice and data communications of public safety workers such as law enforcement officers and national authorities can be protected from being intercepted by individuals who are potential threats to public safety. There can be illicit organizations that can attempt interception of the signals of public safety organizations. Therefore, it is crucial that the over the air information of the all kinds of public safety organizations should be coded or encrypted. Any information flaw may result with intrusions, theft attempts and looting or may affect the performance of public safety missions. Moreover, employment of cognitive radio technology with adaptiveness, learning and decision making features introduces very important, complex and challenging security problems. A cognitive radio may exchange very important information such as local policies, its user's transactions, its user's private communications, etc., the information over the air may be exposed to a malicious attack. Hence, an attack aiming at changing the critical parameters in a policy will definitely cause a communication disaster. For instance, alternating the emission masks may cause tremendous increase at the interference levels observed by the radios which can lead to cease of communication. On the other hand, cognitive radio must support multiple levels of security, furthermore, it is expected from cognitive radio to switch between security levels depending on the situation, application, and environment automatically, within the context of being aware. For instance, cognitive radio must distinguish its actions and deploy appropriate security levels for each of them. As a consequence, cognitive radio must take into account the loss due to the encryption overload [250].

A critical requirement for an encrypted communications system is that it must be able to obtain an encryption update, i.e., to obtain a current encryption key to support secure communications with appropriate destinations and authorized users. Although when the device will be reprogrammed it is historically accomplished by transporting the device to a secure facility, ideally, the new encryption key should be able to obtained remotely by authorized subscribers because the user can be unable to go to a specific location for this operation. Therefore, over the air keying is an important system requirement that enables encryption update using the available radio channels, rather than physical transportation of the equipment [251]. Mentioned requirements of secure communications imply consideration of security issues through the research and design process. The key question is how security of both voice and data communications can be accomplished while taking advantage of flexibility of CR technology.

Tactical Radio System (JTRS) program of the US Department of Defense to standardize the development of SDR technology [252]. SCA establishes an implementation-independent framework with baseline requirements for the development of software for SDRs. The SCA is an architectural framework that was created to maximize portability, interoperability, and reconfigurability of the software while still allowing the flexibility to address domain specific requirements and restrictions [253]. The SCA defines standard interfaces that allow waveform applications to run on multiple hardware sets. The SCA introduces a core framework (providing a standard operating environment) that must be implemented for all SCA capable hardware environments. The Core Operating Environment (COE) consists of a POSIX compliant Operating System (OS) and Common Object Request Broker Architecture (CORBA). Interoperability among radio sets is enhanced because the same base waveform software can be ported to all radio sets [254]. The advancements and gained expertise through the evolution of concepts and architectures such as SCA introduce crucially important input to the design of cognitive public safety radios that are expected to satisfy the introduced requirements.

7.3.4 Location Awareness

Location awareness is a wireless communications feature which is introduced with cognitive radio and cognitive radar concepts. Its purpose is to estimate location of wireless nodes, depending on the extracted information from incoming signals [243–248]. A cognitive radio conceptual model including location awareness engine is introduced in [246]. The proposed location awareness engine is composed of location sensing, location awareness core, and location-aware algorithm adaptation subsystems. Radio sensing methods including range-based schemes, range-free schemes, and pattern-matching based schemes, radio vision methods, and radio hearing methods constitute the location sensing subsystem. The main objective of the location awareness core is to acquire location information, establish relations between these information and make decisions and predictions about the locations of emitting radio (either self or other). Seamless positioning and interoperability, security and privacy, statistical learning and tracking, mobility management, and location-aware applications compose the core. Besides some other applications related to network and transceiver optimization such as location-assisted handover, location-aware applications introduce location based services (LBS) and location-assisted environment identification applications which are expected to give a significant contribution to the design of the cognitive public safety radio introduced herein. Specific implementation of the location awareness concept related to public safety and emergency case requirements should be handled through the location-aware algorithm adaptation subsystem of the location awareness engine. Location awareness for public safety cognitive radios will be considered in the following section.

7.4 Location Aware Cognitive Radio for Public Safety Communications

One of the most effective ways of tracing victims in extreme cases and providing coordination between first responders is to track the wireless signals emitted by the devices in the environment. Recent developments in the wireless communication area have also impacts on emergency call systems such as enhanced 911 (E911) services. These systems are designed to provide improved service for victims especially with respect to instant delivery of the victim's location information to the local Public Safety Answering Point (PSAP) to which the caller is connected. Wireless E911 services have two phases; while the Phase I only asserts the identification of the caller ID and cell phone tower for location estimation, final stage of Phase II requires wireless service providers to provide even more precise location information, specifically, information accurate to the closest PSAP [255]. Wireless service providers employ techniques either based on radiolocation methods depending on wireless network or assisted techniques based on location services such as global positioning system (GPS) over mobile terminals. Network approach can apply triangulation techniques such as angle of arrival (AOA) and time difference of arrival (TDOA) or can be implemented using location signature by implementing fingerprinting methods [256]. Even though the tracking capabilities are improved, network based location information systems and services developed on the requirements of E911 (or E112 in Europe) are insufficient from the aspects of:

- Dependency on the network infrastructure which can be damaged during an extreme situation.
- Congestion due to the capacity limit of either network or PSAP, resulting latency or impossibility of providing victim's location information in timely manner.
- Some factors that are not directly related to the E911 system such as if the victim calls some other number for help instead of 911 and if the batteries of the victims radio is low to make a call.

Even if the original core telecommunications network is down, search and rescue efforts can benefit from prevalence of cellular wireless systems during the extreme cases by communicating with the victims, or their locations can be estimated by extracting the required information from the received signal. Even though there had been some efforts to detect the location of the victims in disaster environments by employing UWB communications and TETRA system jointly, these approaches are limited to the range of a few meters due to the characteristics of the employed technology [227]. However, location awareness feature of cognitive radios can provide a comprehensive solution to the location sensing requirements of public safety communications in extreme cases. It should be aimed to investigate how to accomplish communication with the survivors and how the estimation of their location, utilizing received signals can be achieved [257]. This specific aspect of location awareness requires deployment of public safety radios which combine the channel models developed for extreme cases and physical layer reconfigurability taking the security issues into the consideration. The ad-hoc networking concept can also contribute to the goal; distributed radios can behave like sensors for victim transmission detection and location sensing methods can also benefit from the distributed radios for location estimation.

Detecting and locating victims in a disaster area through their wireless devices and the nature of the mentioned problems above require research for intelligent, secure and robust communication techniques. In this aspect, even though cognitive radio technology is commonly related to public safety communications from the aspect of interoperability, the cognitive radio features of awareness, learning, and intelligence can be beneficial for victim and first responder location estimation. Location awareness feature of CRs aims to provide useful information to wireless devices and networks for enabling them to interact and to learn from surrounding environment [246]. Therefore, location based services of cognitive radio technology can be employed to detect and locate victims.

When the wireless communications systems are considered, location estimation research can be discussed in three folds: (i) channel modeling, (ii) location sensing, and (iii) positioning algorithms [258]. Channel modeling deals with estimation of channel parameters and assessment of the effects of these parameters over the proposed system model. Location sensing feature introduces methods to define and to compute the metrics for location estimation. On the other hand, positioning algorithms are employed to estimate the location coordinates for wireless devices taking the channel and location metrics into consideration. In this section, location awareness feature of cognitive radios will be discussed focusing on the public safety communications from the aspects of the location sensing and positioning algorithms. Proposed approach is also illustrated in Figure 7.3.

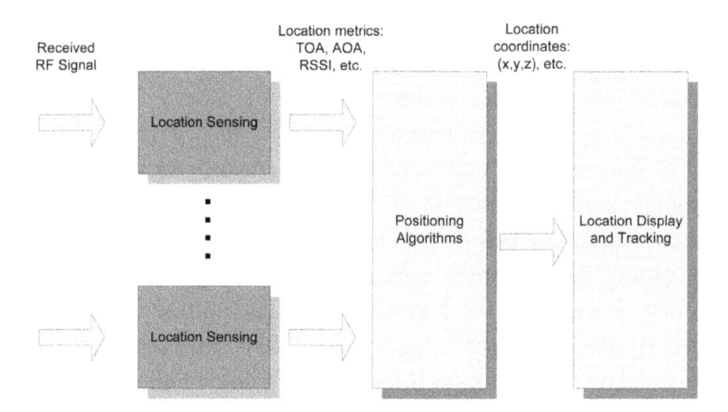

Figure 7.3 Location estimation processes path

7.4.1 Overview of Location Awareness Engine

As the name implies, the "Location Awareness" concept can be defined as being cognizant of location and associated environment. The creatures in the nature have been considered as models for most of the innovations in science history. Similarly, most of the creatures in the nature have already location awareness capabilities into some extent and they have been considered as models for incorporating such capabilities to electronic devices [259]. For instance, bat has location awareness capability, which is known as echolocation, for the navigation and prey capturing [260]. The bats emit high frequency ultrasonic signals (20–200 kHz) from their mouths (transmitter) and listen to the echoes from the environment using their ears (receivers). The received echoes are processed by these animals for different purposes such as navigation, object recognition, and ranging. More intricate example is the human being that is equipped with sophisticated location awareness capabilities. The human beings have multiple sensors such as ears, eyes, and skin that can be utilized for being aware of their locations and corresponding environments. Moreover, the collected signals through these sensors (e.g. optic and acoustic signals) are converted into electrical signals that the brain can interpret. The human being can be aware of its location and surrounding environment by processing the sensed signals in the brain. Consequently, the human being can adapt himself/herself to the environment accordingly. As a result, location awareness mechanisms in the human being mainly consist of sensing, awareness, and adaptation processes is illustrated in Figure 7.4 [246].

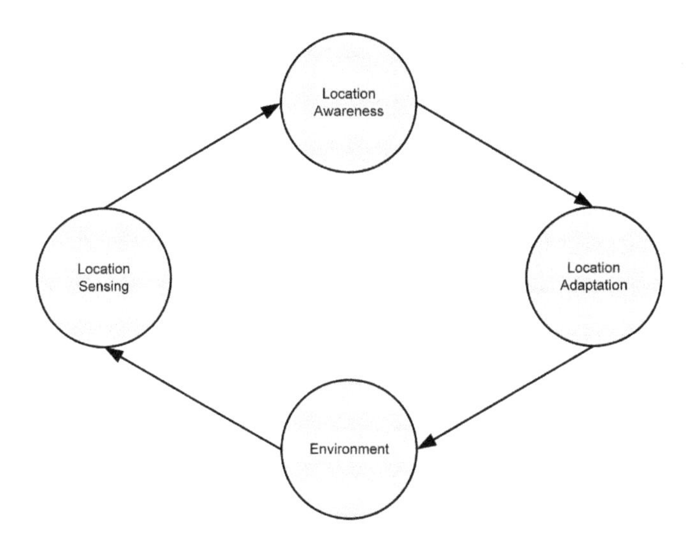

Figure 7.4 Conceptual model of location awareness cycle for creatures [246]

Location awareness can be introduced into electronic systems and such approaches have been investigated extensively for biologically inspired robotics [261]. However, this is not the case for wireless systems. Utilization of location information in wireless systems has been limited to positioning systems and location based services (LBS). Nevertheless, the aforementioned advanced location awareness capabilities of human being or bat can be introduced into wireless systems as well [245, 262]. This can be accomplished by using cognitive radio technology, which was defined previously [206, 263]. According to the definition, cognitive radio has sensing, awareness, and adaptation features, which are the main ingredients of location awareness conceptual model shown in Figure 7.4 [245]. Hence, the consequent conclusion is that cognitive radio is one of the most promising technologies towards realization of location awareness in wireless systems [243, 245, 262].

Location awareness engine receives the tasks from cognitive engine and it reports back the results to the cognitive engine for achieving autonomous location-aware applications at the hand. Furthermore, location awareness engine can utilize various sensors to interact with and learn the radio environments. The proposed model for the location awareness engine in CRs is illustrated in Figure 7.5. According to this model, measurements and/or sensing devices are used to obtain signals from operational environments. The acquired signals are sent to the location awareness core for data post-

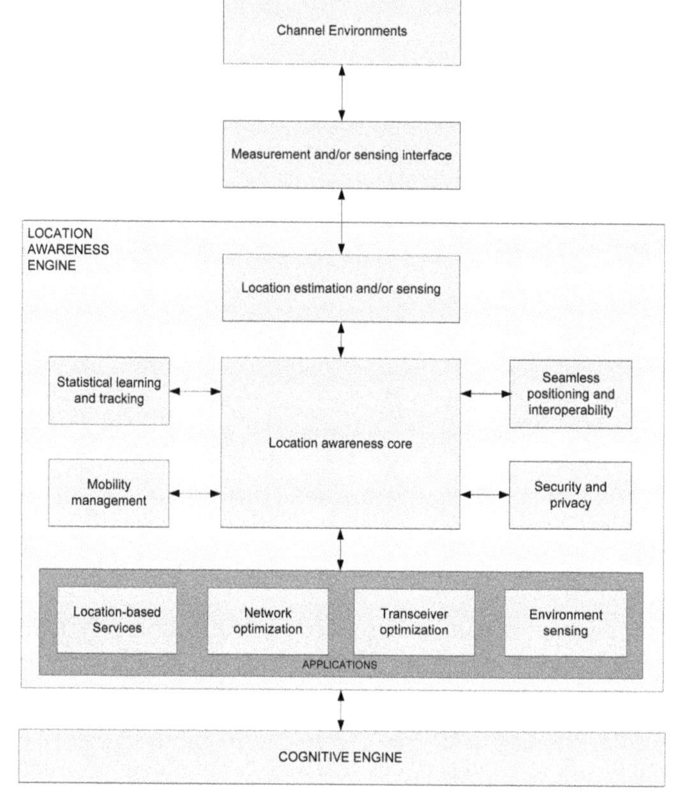

Figure 7.5 Block diagram of location awareness engine for cognitive radio networks [245]

processing. Location estimation and/or sensing algorithms process the signal to determine location information. There are different positioning technologies (e.g., GPS and UWB positioning) as well as reference data types (e.g., WGS84 used by GPS) employed over the world. Therefore, conversion between these technologies and data types is an important issue for public safety radios especially when the first responders are charged with a duty abroad in a country using a different system. CR with the location awareness engine is a promising technology to realize seamless positioning and interoperability between different positioning systems in such cases. The operational block defined as seamless positioning and interoperability in Figure 7.5 has the capability to convert the estimated and/or sensed location information format to other existing reference data or coordinate systems to support interoperability between different positioning systems. Cognitive radio can also benefit from the availability of the services in an adaptive way. For instance, a CR can have the capability to switch from GPS to an indoor positioning system (wireless local area network (WLAN)-based positioning systems) as the user walks from outdoor to indoor environments.

The location awareness engine can have the capability to track mobile CR users and can be trained by the tracking data using statistical learning tools [264] such as neural networks and Markov models to form user location profiles. These profiles can be used to predict the trajectory of CR users and improve positioning accuracy, especially in pattern-matching-based positioning methods. As a result, the location awareness engine can be able to track users with history using statistical learning models, as shown in Figure 7.5. Utilization of location information in CRs and cognitive wireless networks (CWNs) for different applications will have a major impact on system complexity. Introduction of such additional services and applications into CWNs can affect the mobility and quality of service because, the system capacity and implementation cost will change. Therefore, it is desirable to develop an accurate mobility model during the network planning phase. Location awareness engine has a mechanism called as mobility management to handle such tasks, as shown in Figure 7.5.

Extensive utilization of location information in the CRs and CWNs prompts two important issues, which are privacy and security concerns. Some users may not agree on the use of their instantaneous location information for the aforementioned location-awareness-based applications. In such cases, different options need to be available to users to protect their privacy. For instance, in emergency cases such as dialing 911, CWNs should have the right to locate the position of users. On the other hand, providing location information to the network can be optional in non-emergency cases. Another possible solution is that CRs can have self-positioning capability (e.g., a GPS receiver), the network can not determine the user location and users can have the full control over reporting their location to the network anytime they want. Even though these approaches will help with the privacy and security issues, there is a need to develop privacy protection methods and policies. Furthermore, secure positioning systems can be developed to protect users from adversarial attacks (e.g., position and distance spoofing attacks) [265]. By having a privacy protection policy and a secure positioning mechanism along with encryption technologies one can reduce users fear of illegal activity that can threaten their privacy. In summary, a location awareness engine can have a security and privacy mechanism to handle the aforementioned concerns, as illustrated in Figure 7.5.

7.4.2 Location Sensing Interface

Sensing process is composed of mainly two components, which are sensors and associated data post-processing methods. Similar to the creatures in the nature, different sensors can be used in wireless systems for sensing. Sensors are utilized to convert the signals acquired from environment to electrical signals so that cognitive radios can interpret them. The acquired signals can be in different formats such as electromagnetic, optic, and sound. Therefore, sensors can be categorized under three types; electromagnetic, image, and acoustic sensors. Note that the corresponding data post-processing algorithm for each sensing technique is different. Inspiring from the sensing features of the creatures, sensing mechanisms in cognitive radios can be classified under three main categories based on the type of sensors used; radiosensing, radiovision, and radiohearing. Radiosensing is a sensing technique utilizing electromagnetic sensors and the associated post-processing schemes. Similarly, radiovision is a sensing approach using image sensors and the corresponding post-processing schemes. Finally, radiohearing is a sensing method employing acoustic sensors and the associated post-processing schemes.

Image sensors are utilized mainly in passive manner (only receiver) whereas acoustic and electromagnetic sensors are used in active manner (both transmitter and receiver) in wireless systems. Furthermore, image sensors mainly require to point the cognitive radios towards the target direction whereas antennas do not require pointing. In other words, cognitive radio with antennas can continuously interact with channel environment even if it is located in a pocket or bag. Besides, since cognitive radio propose a common sensing interface including different sensors, it can utilize one or combination of the sensors depending on the autonomous task at the hand. For instance, cognitive radio can use both image and acoustic sensors for supporting autonomous location and environment-aware applications similar to utilization of both eyes and ears by human being. Although sensing interface is a common component in cognitive radios to interact with environment and other users, here the focus is on the sensing methods for location awareness systems for public safety communications.

7.4.2.1 Radiosensing Sensors

Antenna is the most widely used radiosensing (electromagnetic) sensor in wireless systems. Therefore the main radiosensing sensor is the antenna which converts electromagnetic signals into electrical signals and vice versa. Physical properties of the received signals and the physical environment that the radio is allocated are the two fore coming deterministic paradigms that define the structure of radiosensing sensors. When the location estimation methodology is considered, most common employed signal statistics are time-of-arrival (TOA), received signal strength indicator (RSSI), and AOA [243]. On the other hand, especially for the public safety communications, the physical environment is defined by the communication medium, which is an important factor affecting the performance of the radiosensing sensors as detailed in Section 7.3.1. For instance, under extreme situations such as storms and hurricanes weather-induced impairments can affect the performance of wireless systems. Under these conditions performance of cognitive radios and networks can be improved by adapting the communication parameters taking the meteorological information of the operating environment into consideration. Wireless channel impairments information for specific extreme cases can be estimated via channel modeling techniques introduced as the first item of the location estimation research in Section 7.4 or such information can be acquired by cognitive radios either from a central server or embedded auxiliary sensors such as thermometer and barometer.

CR technology proposes enhancements to the classical radiosensing applications through the adaptiveness and intelligence features and cognitive radios will have advanced location awareness capabilities. Electromagnetic wave behavior of light could have led to consideration of image sensors in this section however these sensors which uses the light as the source will be studied in the next section due to wide usage of image sensors in the literature.

7.4.2.2 Radiovision Sensors

Radiovision sensors are the devices that acquire the optical signals from the environment and convert them to the electrical signals. These sensors constitute the core hardware of the devices such as digital picture and advanced video recording cameras. Computer vision is a multidisciplinary research field which benefits from these devices and aims to imitate the human vision with the capability of analysis and comprehension of acquired data. Cognitive radios can benefit from the computer vision methods and cognitive vision systems [267] with the capability of scene analysis [262, 266]. Public safety cognitive radios with vision capabilities can quickly record the operational scene through their radiovision sensors, convert the acquired data into an executable format, e.g., text, image and voice, analyze the acquired data and disseminate the gathered information to the public safety officers on the field. This approach would be helpful in case of wide spread disasters such as floods and earth quakes. As the initial visuals are acquired from the field, the scale of the operation to be planned can be decided precisely. Hardware requirements for the crew to be sent to the field can be selected properly. All these actions will lead to significant gain of time in the first vital hours of the incident. On the other hand, if the communications devices of the public safety officers are equipped with radiovision sensors, tracking of the victims and other living things or items which have priority of saving can be detected by pattern recognition techniques. Outputs of multiple radiovision sensors can be combined for further analysis of the scene. Beside the radio systems, another application field for radiovision sensors is the motion detectors which initialize the rapid switching off relays for power and natural gas lines in case of extreme situations such as earthquakes.

One of the main challenges for radiovision applications is assembling such advanced cognitive vision systems to the radio due to low power, cost, and size limitations. Beside that, radiovision sensors have to be positioned towards the target or aimed trajectory. These issues may lead to practical problems in general, however they can be alleviated in the public safety communications. This is mainly because of the facts that the public safety officers' equipments have better battery capacity than regular radios and the uniforms of the officers and first responders provide the opportunity to locate the radiovision sensors over them in an optimal way. For instance, in case of a rescue operation in an urban area spread fire, image sensors (e.g. video camera) along with Ultra-wideband (UWB) transceiver can be mounted to the firefighters hat, which is known as wearable computing devices in the literature [268] and digital camera can acquire the images and then transmits them to the cognitive radio located in one part of the uniform (e.g. pocket) for the data post-processing using UWB transceiver.

7.4.2.3 Radiohearing Sensors

Radiohearing sensors are designed to detect and record acoustic or sound waves, to convert them to the electrical signals and vice versa. Sound wave propagation is used by radiohearing sensors to navigate and detect the objects in an environment while establishing communication between different entities. When the utilization of radiohearing sensors for cognitive radio systems is considered, acoustic location estimation techniques (e.g. sonar [269]) can be employed for cognitive location-aware applications. In this context, cognitive radio can function like a passive sonar such as human ear or behave like an active sonar such as bat echolocation in nature depending on the application. For instance, one major application for CR with passive sonar type of radiohearing sensor is indoor detection and tracking of human beings and other living things. Indoor location estimation and tracing is vital for various extreme cases such as rescue operations from the piles of blocks in earthquakes, trapped victim detection in case of fire in a building, and indoor criminal tracing.

One way to utilize active acoustic sensors in cognitive radios can be the sensing of the surrounding environment from the acoustic waves transmitted and received, similar to human beings and bats. In extreme cases if the visibility is limited, save and rescue operations personnel can benefit from radiohearing sensors to identify the nature of the operation environment. Beside that, the microphone which is integrated to the cognitive radio can be used as a passive sonar to record environmental sounds. The captured signals can be compared to the sounds stored in a database and some certain environmental features can be extracted or changes in an environment can be detected by comparing the most recent recording with the previous ones.

7.4.3 Location Sensing Methods

7.4.3.1 Radiosensing Methods

Radiosensing methods are antenna-based location sensing applications which can be categorized under three groups: range-based schemes, range-free schemes, and pattern matching-based schemes. In range-based schemes, a set of parameters are used to estimate the location information of the target device. These parameters can be TOA, RSSI, AOA, and TDOA for classical radiosensing methods. The statistical information acquired from the received signal are used to calculate these parameters. On the other hand, range-free schemes are simple sensing methods which provide a coarse location information. Hop-count-based approach (e.g. ad-hoc positioning system (APS)), centroid, and area based approach are the three main approaches for range-free schemes. CRs can benefit from the range-free schemes especially over the wide spread operation areas such as wild forest fires and large scale floods for estimation of the location of the victims in a coarse way. A couple of these methods, such as APS and centroid can be combined adaptively for better estimation results. Pattern matching-based schemes are based on probabilistic models and they can be used to estimate the terminal locations when signal measurements are available. Various attributable variables can be used such as signal strength and power delay profiles and different probabilistic modeling approaches can be employed based on the type of the application. For instance, Bayesian and Hidden Markov models are frequently used for indoor location sensing applications. Hence, pattern matching-based schemes can be deployed by public safety cognitive radios for the cases such as fire rescue victim locating and short range tracking.

Range-based schemes provide a wider set of parameters and have extensive applicability when compared to the two other schemes. Therefore, some more detail will be given about these schemes; applications and opportunities related to cognitive radio will be discussed in the context of range-based schemes. Proximity and triangulation are two main range-based location estimation techniques. Proximity is a location estimation method which depends on the received power levels over the antenna of the mobile device. It is a standard or technology dependent functionality which can be employed over centralized networks such as cellular systems and WLANs. Each cell or wireless node has a unique identifier for its location, e.g., Location Area Identifier (LAI) which is broadcasted with the cell identifier. Therefore, mobile station can determine its location relatively to the base station or wireless node. The accuracy of the location information depends on the cell or coverage area size [270]. Public safety cognitive radios can benefit from proximity based algorithms when the wireless infrastructure is not damaged and available. CR can identify the existing network in the operation area and adapt itself to it for location estimation. Demodulation of the signals such as over the control channels can be required for precise location estimation, which is a procedure that may increase the location estimation time.

Triangulation depends on the geometric properties of triangles to estimate the location of a device. Angulation (i.e., AOA) and lateration (i.e., TOA, RSSI) are two forecoming triangulation techniques. While angulation methods utilize both angle and distance for positioning, lateration methods employ distance to triangulate the location of the device to be traced. When the AOA estimation is considered, the angles of reference objects are measured with relative to a reference vector which is defined for all reference points. The dimensions of the operation space defines the number of reference points and their geometric relationships. For instance, angle information from two different reference devices and the distance information between both reference points are needed to estimate the location of a target point in two dimensions [271].

The number of independent points, i.e., the number of the independent equations required to solve will change based on the number of spatial dimensions, e.g., 1D, 2D, 3D for lateration, because lateration is applied by measuring the distances of independent reference points with regarding to the target device. RSSI is simpler than TOA estimation, however, its estimation resolution is lower when compared to TOA in general. TOA algorithms compute the transmission time between the transmitter and receiver as the name implies. The estimate of the distance information is extracted from the wave travel time and the wave velocity. CR can employ different waveforms such as sound and light based on the environmental and other parameters. For instance, acoustic waveforms can be used for underwater rescue operations. The GPS utilizes the concept of one-way TOA ranging to estimate the position.

Triangulation and proximity require a certain amount of measurements based on the technique used. The type of location information that needs to be estimated or sensed plays an important role to determine the complexity of the positioning systems. Location information is categorized as absolute and relative location information. Absolute location refers to complete set of coordinates that give the location of the target on the predefined set of measures such as North American Datum (NAD) and World Geodetic System (WGS). On the other hand, relative location information implies that the position of the target relatively to another reference point that the absolute location is unknown [272]. In the majority of location estimation and sensing techniques, the dimension of location information that needs to be estimated or sensed determines minimum number of reference devices required and the geometric relationship between them. For instance, distance measurements from three devices (multi-lateration) that are located in a non-collinear manner are required to estimate the location of a device in 2-D. On the other hand, estimation of the 3-D location of a device requires the distance measurements from four non-coplanar devices [273]. In general, the number of reference locations should be one more than the dimension of the search space and public safety applications for CR can benefit from public safety officers' radios located on different places in the operation environment to satisfy the reference point requirements. Each radio used by the public safety officers can be defined as an independent node or as a reference point and the optimal location estimation method can be employed by operators in the central command center who can communicate with the first responders. Moreover, public safety cognitive radios can configure themselves automatically using the information which will be provided by the location awareness engines [246]. As a result, CR can optimize the performance and complexity of the positioning algorithm by having a priori information about the dimension of the location information.

Legacy location estimation techniques such as TOA, signal strength, and AOA can be considered as the candidates for the cognitive positioning system (CPS) of public safety CR. AOA techniques are mostly implemented by means of antenna arrays. But, angulation employing antenna arrays is not suitable for rich multipath environments such as indoor propagation channels due to the cost and imprecise location estimation caused by indistinctive multipath components [274]. On the other hand, signal strength based methods provide high accuracy only for the short ranges since the Cramer-Rao Lower Bound (CRLB) which define the upper estimation limit for these methods depends on the distance inversely [275]. Moreover, the performance of the estimator for signal strength techniques depends on the channel parameters such as path loss factor and standard deviation of the shadowing effects which are difficult to track precisely due to their rapidly changing nature. Therefore, the accuracy of signal strength techniques can vary because CR does not have any control over the channel parameters. Finally, as indicated above, accuracy of TOA techniques mainly depends on the parameters, e.g., number of reference points, that transceiver can control. Therefore TOA can be considered as the most suitable location estimation technique for the public safety applications.

The accuracy and complexity of the employed positioning technique can affect the performance of the location awareness related applications. GPS and UWB are the two existing positioning technologies. There are different forms of GPS technology; standard GPS (4–20 m accuracy), Code-Phase GPS (3–6 m accuracy), Carrier-Phase GPS (3–4 mm accuracy), Differential GPS (sub-decimeter) [276], Assisted-GPS (less than 10 m accuracy) [277], Indoor GPS (1–2 cm accuracy) [278], and Software GPS (less than 15 m accuracy) [279]. Each of these GPS technologies provide different level of accuracy and combining these different forms of GPS in a single device is impractical and costly for public safety applications. However, software GPS is a promising method to switch between different GPS forms. But, eventually, this approach will only provide a set of fixed accuracy levels that are provided by each form of GPS. Basically, the existing GPS technologies do not have a capability to achieve accuracy adaptation. Another alternative technology is UWB positioning, which has the capability to provide centimeter ranging accuracy due to the use of large bandwidth during the transmission. However, this technology does not have a capability to achieve accuracy adaptation for positioning either. Moreover, this technology provides such fixed and high-precision positioning accuracy only within short ranges. Even tough they do not provide the level of adaptiveness, these legacy technologies can be employed within the location aware public safety CR. In these cases, GPS applications will require a certain number of satellites to cover the mobile stations operation area and this condition may not be satisfied for some events such as disasters spread over wide areas. Beside that, in general GPS is used as a self-positioning system which the radio estimates its own location. Therefore, in some cases it can be difficult to implement it for victim tracking. CR can configure itself to UWB mode for penetrating through obstacles but, range will be limited to a couple of meters as indicated before. Taking all these facts into consideration, legacy positioning techniques without enhancements do not provide the required cognition capability that public safety CR demands. However, radiosensing methods can benefit from the existing positioning technologies whenever they are available by combining them with the CPS to improve the accuracy of location estimation.

One of the properties of CRs is that they can adjust the assignment of resources depending on the application. Location awareness applications of public safety communications may require different levels of positioning accuracy. For instance, precision of the position information can be expected to change if the operation environment is indoor, indoor/outdoor combination, and extremely wide areas. CR can automatically select the optimum set of parameters independently of the radiosensing method used. In case of extreme cases, because the time is the most critical parameter, intelligence and adaptiveness features of CRs are helpful for first responders. For instance, when they reach to the operation area, first responders do not have to take care of the adjustments of their devices. Instead, they can focus on their missions in a more effective way. The aim of the proposed location awareness feature is to carry the current positioning methods and technologies to the next step. Therefore, instead of focusing on the technologies the scope of the research covers the methods and algorithms such as TOA, AOA, and RSSI which are behind the technology. In this aspect, dynamic spectrum management (DSM) in CR technology can be used for both communications [280] and positioning systems.

The performance and optimization requirements for communications and positioning will be different. For instance, one of the main performance parameters in the communications systems is data rate, whereas it is accuracy in the positioning systems. Similarly, the optimization algorithm that is used by DSM for the communications and positioning systems can be different. The optimization algorithm used by DSM to support positioning systems is referred as enhanced dynamic spectrum management (EDSM) [243]. The adaptation of positioning accuracy is achieved in two steps: bandwidth determination and EDSM. In the bandwidth determination mode, location awareness core determines the required effective bandwidth for a given accuracy. The required effective bandwidth can be determined using the bandwidth determination equation, which should be derived through CRLB for both additive white Gaussian noise (AWGN) and multipath channels for extreme cases which are detailed in Section 7.3.1. In practice this is done by the estimation of path delay coefficients and noise variance using log-likelihood function and by calculating variance of delay estimation error [243]. Once the effective bandwidth is determined, the second mode, that is, the EDSM system is initiated. The main responsibility of the EDSM is to search, find and provide the optimum available bandwidth for location awareness core to be able to satisfy the positioning accuracy requirements. Finally, the specified relative bandwidth is used by one of the reference CR to transmit signal, and selected positioning algorithm, e.g., an adaptive TOA based location estimation algorithm (A-TOA) which has the range accuracy adaptation capability is employed by the target CR node to estimate the location with given accuracy. Note that it is assumed that the reference and target CR nodes agree on the relative bandwidth during the initial ranging handshake mechanism. Adaptive positioning algorithms can be deployed by public safety CRs in an effective manner due to the fact that in general there will be at least a couple of public safety officers and a couple of CRs in the operational environment which will lead to cooperation between radios. Beside the target signal detection and tracking, positioning algorithms can also be used to monitor the locations of the first responders during the operation. This helps with the optimization of operational distribution of the first responders over the field and with the monitoring of the progress of the events on the field. Positioning techniques can also be classified into two categories depending on the choice of implementation: (i) decentralized (self)-positioning, (ii) centralized (remote)-positioning. These approaches are illustrated for public safety CRs from the aspect of victim tracking in Figure 7.6.[1]

7.4.3.2 Radiovision Methods

Image sensors which can be located as wearable computing devices can produce visual data such as video camera signals (e.g., video). These data related to the scene can be acquired and sent to cognitive radio to be processed using advanced digital image and signal processing techniques (e.g. pattern analysis and machine intelligence algorithms such as dynamic 3-D scene analysis [267]). The information that is extracted through the processing methods can be stored in different formats such as text, image, video, and voice. The storing format can be selected based on the storage capabilities

[1] PSCR: Public Safety Cognitive Radio.

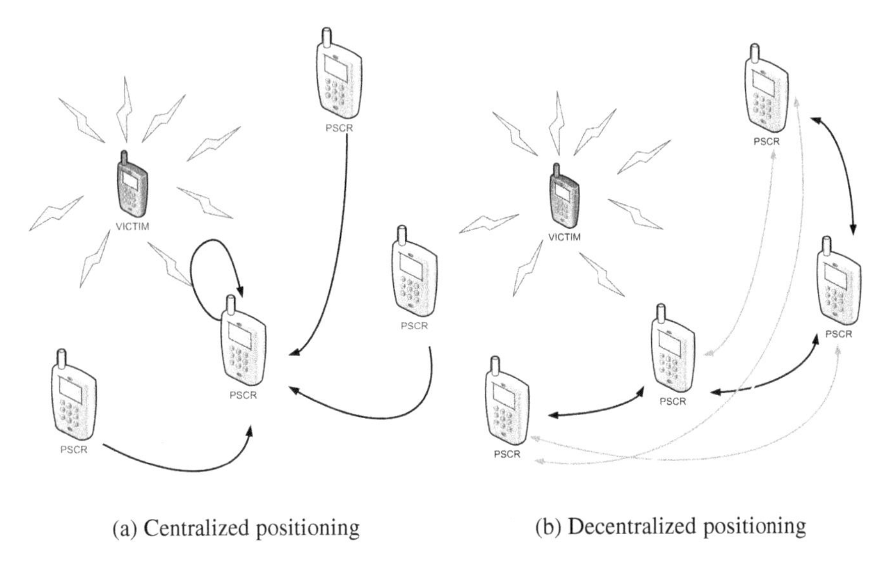

(a) Centralized positioning (b) Decentralized positioning

Figure 7.6 Implementations of positioning techniques

and some other requirements such as scene reconstruction fidelity. Scene analysis [267,281,282] is location sensing technique which is based on detection of previously known patterns in an environment. It is similar to RF pattern matching based methods, i.e., fingerprinting and match filtering [283] and can be employed as a visual location sensing technique in extreme cases. Acquired images can be searched to detect patterns in the scene analysis, which is also similar to the utilization of the wireless channel statistics (e.g. TOA) in the RF pattern matching based methods.

Instead of a precise location information, radiovision based location sensing methods can provide location information with less resolution when compared to radiosensing methods. A better location accuracy can be achieved by computing the relative distances of the detected objects in a given scene. Therefore, radiovision based sensing methods can be preferred for object and environment recognition rather than location sensing. However, such an approach requires a database of the potential objects that can be encountered during public safety operations and an extensive computational power for the execution of radiovision methods. Although these drawbacks do not favor radiovision methods as a general application for location sensing, algorithms based on scene analysis have an important application potential for public safety communications. If the radiovision sensors are attached to the uniforms of the first responders, especially for the wide spread disasters such as wild forest fires and extensive floods it can be possible to estimate the size of the event which will lead to selection of proper items to be sent to the field and realistic planning.

Implementing radiovision techniques such as cognitive vision systems in cognitive radios is a challenging task due to low power, cost, and size constraints. However, these problems can also be alleviated in the case of public safety communications. Again, taking the fact that the number of the CRs in the field will be quite high, these operations can be completed in a collaborative manner. CR can schedule the processing times depending on the communication schedule of the responders. During the silence period, the extensive image processing work can be distributed and handled in parallel by the CR devices in the field. An approach similar to centralized positioning methods as depicted in Figure 7.6a can be implemented.

7.4.3.3 Radiohearing Methods

The acoustic sensors or the devices to record the sound waves provide the data that is used by radiohearing based location sensing methods. Beside that some sound generators can be utilized by CR for interacting with environment if the device is designed as active sonar. Radiohearing based location sensing methods can be categorized into same group of schemes which was introduced for radiosensing methods: range-based, range-free, and pattern matching based techniques. The first two of these methods are studied more when compared to pattern matching based techniques, especially for legacy location estimation techniques. Cognitive radio is a platform with adaptiveness and intelligence features which advanced radiohearing based location sensing techniques can be realized upon. In this context, bat echolocation can be simulated as an active sonar and CR can transmit a sound wave and listen the echoes of it. It can compare the acquired sound signals with a set of signals stored to identify a pattern or the spectrum of the captured sound waves can be compared with some known spectral patterns for location sensing.

One of the implementation of radiohearing methods is sonar like CRs [246], which can lead to the estimation of the location of the victims either lost in a disaster area or stuck under the pile of ruins by employing A-TOA location estimation methods based on the sound waves. Radiohearing methods can also lead to detection of irregularities of the acoustic waves in an environment and can help to early detection of environmental occasions such as wild forest fires and rapid floods before they spread over wider areas. In an extreme case, the initial time periods are critical to keep the damage and size of the occasion limited. Therefore, early detection leads to earlier response of the first responders and to the initialization of warning systems for civilians before the disaster is spread over urban and suburban areas.

7.5 Application: A Signal Detection and Power Estimation Method for RSSI-Based Location Sensing

For the RSSI based positioning algorithms, the CRLB which is derived for position estimation calculations, is a function of standard deviation of shadowing, path loss exponent, and distance. In practice, the received signal power is calculated and then the path loss power is obtained assuming that the transmit power is known. Later on, the distance is estimated using the path loss equation. In this section a peak power estimation algorithm based on noise floor estimation will be introduced for a RSSI-based location sensing system. The output of the proposed method will provide the received signal power input to the positioning algorithms illustrated in Figure 7.3.

Decision process of the location sensing metrics has vital importance in the aspect of positioning performance of location estimation algorithms. Accuracy, robustness and complexity are major criteria for effective and fast location sensing. Taking these measures into consideration, the main goal can be defined as introducing a signal detection algorithm that will work fast and give robust results in extreme situations. For this reason, the application case should be defined as realistic as possible. Therefore, instead of focusing on a single communication channel, it is assumed that the wireless equipment is able to sweep a portion of the frequency spectrum and there is no prior information about the occupant signal(s). It is also assumed that the spectrum can be swept any time of any sampling rate with a resolution bandwidth that will not lead to the loss of spectral information. Taking these information into consideration, transmit signal model can be given by

$$x(t) = As(t), \tag{7.1}$$

where $s(t) = [s_1(t), \ldots, s_n(t)]^T$ represents n independent signals and T denotes transposition. A is $n \times n$ coefficient matrix with $a_{i,j}$ elements where $i, j = 1, \ldots, n$ and $x(t)$ represents n transmit signals where $x_j(t) = \sum_{i=1}^{n} a_{i,j} s_j(t)$. After the digital conversion and modulation, each independent signal is assumed to pass through a different wireless channel. The channel for $x_j(t)$ can be modeled as a time-variant linear filter

$$h_j(t) = \sum_{i=1}^{L_j} h_i(t)\delta(t - \tau_i), \tag{7.2}$$

where L_j is the number of taps for the channel h_j and τ_i is delay for each tap. It is assumed that the taps are sample spaced and the channel is constant for a symbol but time-varying across multiple symbols. The transmitted signal is received along with noise at the receiver. Therefore, baseband model of a single received signal, after down conversion can be given as

$$y_j(t) = e^{2\pi j \xi t}[x_j(t) \star h_j(t)] + w(t) \tag{7.3}$$

$$= e^{2\pi j \xi t} \int x_j(\tau) h_j(t - \tau) d\tau + w(t), \tag{7.4}$$

where $\star$ is the convolution operation and the j at the exponential term is the imaginary unit. When the channel model given in (7.2) is considered, the received signal becomes

$$y_j(t) = e^{2\pi j \xi t} \sum_{i=1}^{L_j} x_j(t - \tau_i) h_i(t) + w(t), \tag{7.5}$$

where ξ is the frequency offset due to inaccurate frequency synchronization and $w(t)$ corresponds to additive white Gaussian noise (AWGN) sample with zero mean and variance of σ_w^2. The composite received signal then becomes

$$y(t) = \sum_{j=1}^{n} y_j(t). \tag{7.6}$$

After the received signal is sampled with a sampling time of Δt, the discrete-time received signal can be represented in vector notation by

$$y(n) = [y(1), y(2), \ldots, y(N)]. \tag{7.7}$$

where $n = 1, \ldots, N$. The frequency domain representation of the received signal is acquired by applying fast Fourier transform (FFT) by

$$Y(u) = \sum_{u=1}^{N} y(n) e^{-j2\pi u \frac{n}{N}} \tag{7.8}$$

where $u = 1, \ldots, N$. When the focus is on the parameters in hand, it is known that there are N spectral samples out of FFT and the sequence is assumed to be stationary hence all the samples belong to the same sweep of the frequency spectrum from the selected starting point until the stop frequency. Besides this, when the practical cases are considered, the dynamic range of the recording equipment will define the possible maxima and minima of the signal power that can be recorded by the device. Therefore, at any point in the frequency spectrum if the power level is lower than the minimum power that can be recorded, it will be represented with the equipments dynamic range minima. Hence, the signal power of the frequency bands with no transmission will be represented with the samples that are closely distributed around dynamic range minima because the actual power distribution at these frequencies are either blow or around the dynamic range minima. On the other hand, when the frequency spectrum is considered, the samples with the least power will carry the least information while the samples with the highest power levels will carry most of the information about the wireless occupant signals. Therefore, it is plausible to divide the signal samples into two groups: one group starting from the bottom of the spectra until a level of power that should constitute the signal base or noise floor and the second group of samples should be evaluated as the information bearing samples for wireless signals. This approach is considerably different from applying a threshold for spectrum sensing where the threshold can be selected arbitrarily, e.g., mean value of the frequency spectrum because such a threshold can miss some of the signals with low signal to noise ratio (SNR) in a given spectrum easily, but estimation of noise floor in the first place eliminates such cases.

Accurate noise floor estimation is an extremely important requirement for the calculation of RSSI value in severe environments of extreme cases. Considering the two groups of samples defined, a changing point should be detected in the power distribution to distinguish the noise floor and information bearing samples. The power levels of the bands with no transmission will be represented with the samples that exhibit a denser distribution around the dynamic range minima when compared to the information bearing samples which will be relatively sparsely distributed compared to the first group of samples in the spectrum due to their abilities to represent the actual power values in the signal transmission bands. The denser distribution means lower standard deviation, therefore starting from the minima of the power level, detecting a significant increase in the standard deviation will mark the end of noise floor for a given frequency spectrum. However, there are two issues that should be considered: (i) how to implement decision making process in general, (ii) requirement of a threshold for deciding that the change in the standard deviation is *significant.* First problem can be handled by adopting quantization level approach from the analog signal digitization process: the frequency spectrum will be partitioned into horizontal segments and all the samples will be grouped depending on their corresponding segments. Therefore, if the power spectral representation of received signal is defined as $Y(u)$, where $u = 1, \ldots, N$ in (7.8), the spectrum can be divided into horizontal segments with the length of:

$$L = \frac{\max(Y) - \min(Y)}{k\sigma_{Y(N)}} \tag{7.9}$$

where k is a coefficient and can be selected as $0 < k \leq 1$, and $\sigma_{Y(N)}$ is the standard deviation of whole power spectrum which is defined by:

$$\sigma_{Y(N)} = \sqrt{\frac{1}{N}\sum_{p=1}^{N}(Y(p) - \bar{Y})^2} \tag{7.10}$$

It should be noted that $\bar{Y}$ is mean of the power spectrum and the segment length changes adaptively with the general standard deviation of the recorded spectrum. If the standard deviation is high there are more fluctuation in the frequency spectrum, then length of the segments become narrower to catch the activity. If the standard deviation is smaller then spectrum is relatively flat, segment length becomes wider and a significant gain in terms of complexity is achieved. It should also be noted that $\sigma_{Y(N)}$ can be selected as the detection threshold due to the fact that the general distribution of the multi-band wireless spectrum will always be sparser when compared to the distribution of the samples of noise floor aroud the dynamic range minima: $\sigma_{Y(N)} \geq \max(\sigma_{L_j})$ where σ_{L_j} represents the standard deviations of the samples belonging to the segments constituting the noise floor. Therefore, starting from the bottom level, each horizontal chunk that satisfies: $\sigma_{L_i} < \sigma_{Y(N)}$ will constitute the noise floor, where σ_{L_i} represents the standard deviation of samples of all segments where $i = 1, \ldots, \lfloor k\sigma_{Y(N)} \rfloor$, $\lfloor k\sigma_{Y(N)} \rfloor \leq k\sigma_{Y(N)}$ and $\lfloor k\sigma_{Y(N)} \rfloor \in Z$.

A simulation setup is constructed to test the proposed techniques over the simulated signals. Wireless signals of various bandwidths and power levels are generated. Based on the signal model introduced in (7.5), in all of the simulations noise is present and assumed to be of AWGN. The impact of multipath channel is simulated according to the ITU-R M.1225 outdoor to indoor A channel model [241]. Proposed detection algorithm is compared with the

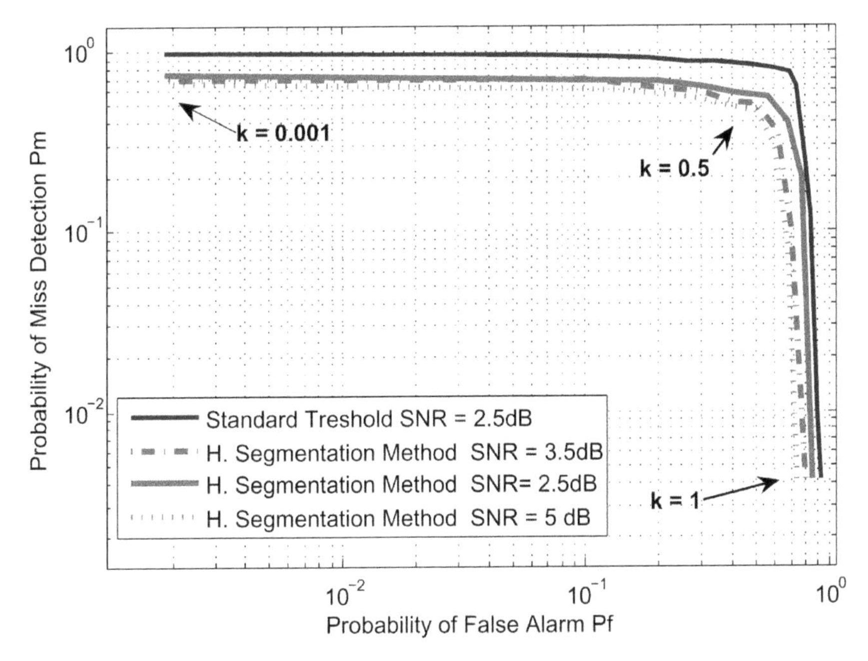

Figure 7.7 Probability of miss detection/false alarm rate for proposed method vs. energy detection

energy detection technique that is proposed for spectrum sensing in cognitive radios [208] considering various k coefficient values for SNR of 2.5, 3.5 and 5 dB in Figure 7.7. Noise floor estimation method has lower false alarm and miss detection rates when compared to energy detection technique for the same SNR level of 2.5 dB. This improvement is due to the segmentation of spectrum and estimation of noise floor. Figure 7.7 shows less than 5% missed detection and false alarm rates for the given SNR levels when the coefficient k is selected around 0.5.

The second part of the proposed method is the estimation of the signal power level. The noise floor detection stage completed the distinction of information bearing samples from the estimated noise floor. In the next step, information bearing samples for each signal are grouped together and the peak power of each signal is computed by selecting the local power level maxima. The performance of the power level estimation algorithm is illustrated in Figure 7.8. The vertical scale represents the normalized peak power estimation error while the horizontal scale is the SNR for the simulated signals. The estimation errors imply that the algorithm is SNR dependent

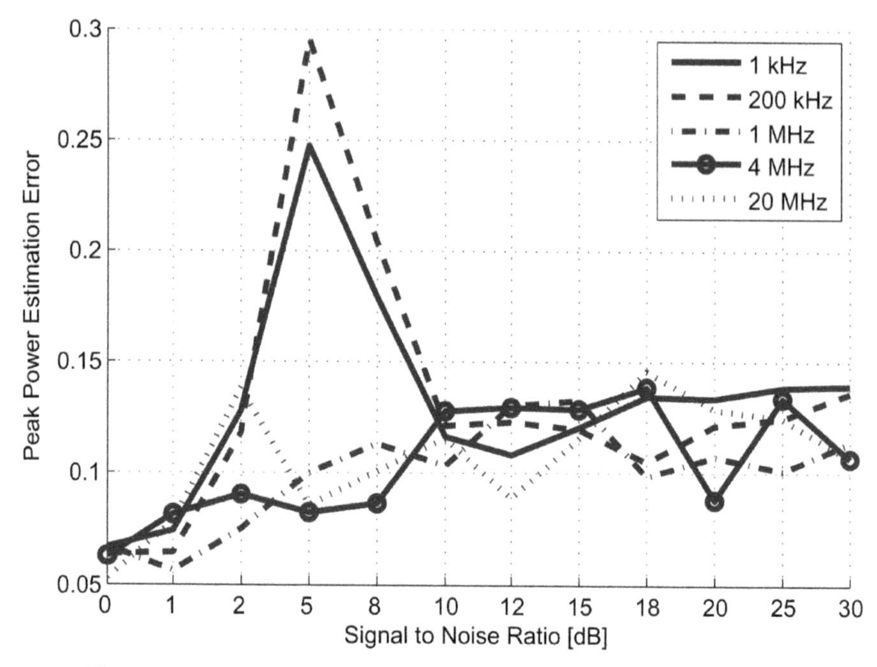

Figure 7.8 Power level estimation error for signals with various bandwidths

and for the low SNR values which are below 5 dB, the algorithm can not produce consistent results. This stems from the fact that for the signals with very low SNR, the distribution of the samples that constitute the noise floor and information bearing samples that constitute the wireless signals become similar therefore, it becomes harder to detect the changing point. The estimation results become consistent for the signals with 5 dB SNR or higher and the error margin drops back to 15–10% band. Peak power estimation is conducted for the signals with various bandwidths, i.e., 1 kHz, 200 kHz, 1 MHz, 4 MHz, and 20 MHz to investigate the dependency on the frequency span. The results in Figure 7.8 indicate that the proposed estimation algorithm is bandwidth independent because the error rates for the signals with different bandwidths are in a close margin for the same SNR values higher than 8 dB. This is due to the adaptive segmentation of the wireless signals which is introduced by (7.9) and (7.10).

7.6 Conclusions

The advanced flexibility, reconfigurability, learning and awareness features of cognitive radios itself forward to solve the current problems of public safety communications systems. Moreover, the assessment of communications requirements in extreme and emergency cases, taking the advancements that can be achieved by the employment of cognitive radio technologies into account, clearly indicates that it can be possible to achieve interoperability and non-ceased communications between the first responders in an efficient, reliable and secure way.

Location and environment awareness features of cognitive radios can provide improvement at the quality of service of public safety communications with the aid of mobility management and statistical learning and tracking techniques. Communication and data storing security can be enhanced by the security and privacy methods of location awareness engine. Moreover, methods of location sensing, positioning and channel modeling brings opportunities from the aspect of location estimation methodology. Most importantly, implementation of location awareness methodologies which leads to adaptive positioning accuracy via the algorithms such as A-TOA can help the estimation of locations of victims in various situations, e.g. over vast disaster areas or under collapsed buildings and improve the organization of the first responders between themselves. The adaptive positioning algorithms and applications based on them works even if the communications network is down in the operation area. Beside the adaptation, reliability is the most important feature of location aware public safety cognitive radios.

Acknowledgment

This work is supported by Qatar National Research Fund (QNRF) grant through National Priority Research Program (NPRP) No. 08-152-2-043. QNRF is an initiative of Qatar Foundation, Doha Qatar.

8

Networks of Mobile Robots for Rescue Operations

Hisayoshi Sugiyama

Osaka City University, Japan; e-mail: sugi@info.eng.osaka-cu.ac.jp

8.1 Introduction

Activities of mobile robots in disaster areas and their abilities to detect victims became known to the public in occasion of the World Trade Center (WTC) disaster on September 11, 2001 [284]. Within six hours from the disaster, four robot teams were dispatched from the Center for Robot-Assisted Search and Rescue (CRASAR) and were applied for the essential tasks of rescue operations under the supervisions of scientists.

Though the robot teams could not detect survivors finally, this tragedy promoted the research activities for urban search and rescue (USAR) utilizing autonomous mobile robots [285, 286]. USAR aims to search for and rescue victims in disaster areas, especially those littered with debris from man-made objects like the area of WTC disaster. Many aspects of the related robotics have been investigated [287] for movement over rough terrain [288], localization and navigation with obstacle avoidance [289, 290], sensor systems that react to human bodies [291], and so on.

In addition to these aspects of single robotics, a multi-robot system and its integrated operations may become essential in the practical design of rescue systems because of its system reliability and scalability. Reliability means that a failure of some robots, which tends to happen especially in hazardous environments, does not seriously affect the system performance itself. Scalability means that the system is applicable to a wide variety of disaster areas with a suitable number of robots supervised by a small number of operators.

N. Marchetti (ed.), Telecommunications in Disaster Areas, 213–241.
© 2010 *River Publishers. All rights reserved.*

The multi-robot rescue system consists of a base station (BS) and multiple mobile robots. All of them communicate with each other through a wireless network. The BS is established within a safety zone around the disaster area, whereas robots walk around the disaster area to detect victims located within the range of their sensors. Operators at the BS monitor the sensed data transmitted from robots and control them, in order to decide if successive rescue operations are necessary.

Although a star-shaped network with a radio center station is sometimes considered as a communication setup for a multi-robot rescue system [292, 293], an ad hoc one may be more suitable for practical operation of a rescue system as investigated in recent works [294, 295]. There are following two reasons for this. First, the radio center station may be difficult to establish at the appropriate point in the line of sight of every robot. The robot may be located at various positions among debris and obstacles or hidden in cave-like spaces under these objects. Whereas, in ad hoc networks, communication paths possibly connect every robot and the BS threading through the objects and no center station is necessary. Second, concerning the limited power source of each robot, its transmission power may be so small that the transmission radius merely includes neighboring robots and not the distant radio station. This limitation of transmission radius is possible to be handled in ad hoc networking.

In this chapter, it is described how ad hoc networking works in multi-robot rescue systems concerning two aspects of their operations. First, after the explanation of the fundamentals of ad hoc networking in Section 8.2, autonomous chain network formations by rescue robots are introduced in Section 8.3. The chain networks are essential for reconnaissance into distant spaces within collapsed buildings or so where victims may be trapped. Second, in Section 8.4, wireless quality of service (QoS) networks for multi-robot systems are introduced. This QoS scheme reserves the transmission bandwidth when a robot sends wideband signals to the BS through the chain network. This scheme is essential in ad hoc networking when robots detect victims and their dynamic picture images must be transmitted simultaneously without packet collisions or delays. These aspects of the networks of multi-robot systems for rescue operations are summarized in the final section.

8.2 Fundamentals of Ad Hoc Networks

An ad hoc network consists of mobile terminals communicating with each other through a preassinged radio channel. Each communication is done by a radio communication link directly or through a communication path indirectly. The communication link (from now on, referred to as *link*) is established between a pair of terminals within transmission ranges of each other. The transmission range is defined around a terminal wherein any other terminal can transmit (or receive) signals directly to (from) the center terminal (this terminal within the transmission range is called the *neighboring one* of the center terminal). On the other hand, the communication path (from now on, referred to as *path*) is established by tracing a cascade of links of intermediate terminals. These terminals relay the transmission of packets until they reach the destination. This process is assimilated to *hops* of packets. *One hop* means the direct transmission of packets through a link and the number of links a path consists of is denoted by the *hop count* of the path.

The path transforms ad hoc when network topology changes along with movements of terminals. The establishments and transformations of paths are accomplished through the collaborative process of every terminal that individually follows a predetermined routing protocol. No particular terminal or center station is responsible for planning the path and specifying the relaying terminals. This decentralized scheme assures the reliability and scalability of ad hoc networking. Similarly to the advantages of multi-robot system described in Section 8.1, the reliability means that a breakage of some link included in a path may not affect the communication of terminals that are temporary using that path. In this case, the path transformation immediately retrieves the communication. The scalability means that basically no restrictions exist on terminal population and on network configuration. Moreover, any incidental terminal additions or subtractions are acceptable to the ad hoc networking.

In the following subsections, first, the classification of ad hoc networks is explained and typical schemes of each class are listed. Among the schemes, DSDV (Destination Sequenced Distance Vector) routing protocol is selected as a basic one and is explained in successive subsections.

8.2.1 Classification of Ad Hoc Networks

As mentioned before, ad hoc networking is accomplished depending on the routing protocol which each terminal follows. The routing protocol mainly specifies how the routing table of each terminal is constructed and updated during the networking. Though many protocols exist in relation with the routing table processing, ad hoc networks are classified into two classes: proactive and reactive routing-based, according to when a path is established based on the routing protocol. In proactive routing, the routing table is updated periodically by *update packets* exchanged among terminals. Therefore, anytime the path is established between any pair of terminals. Destination Sequenced Distance Vector (DSDV) [296] and Cluster Switch Gateway Routing (CSGR) [297] are typical schemes belong to this class. On the other hand, in reactive routing, a path is established when a terminal intends to transmit data packets to another one. Routing tables of relaying terminals are updated coincidentally with this event of calling. This scheme alleviates the congestion associated with update packets, especially in a large scale network or when the network topology changes frequently. Typical protocols that belong to reactive routing include Ad Hoc On-Demand Distance Vector Routing (AODV) [298] and Dynamic Source Routing (DSR) [299].

Among these classes, proactive routing is suitable for the networks of multi-robot systems. This is because in the operations of the system, each robot must always be aware of the current properties of other robots and the update packets exchanged periodically in proactive routing are convenient to subsidiarily transmit the information.

For example, as explained in the next section, every robot must identify the current locations of other ones in the formation and transformation of a chain network. For the purpose of this identification, position information of robots are transmitted periodically with update packets exchanged among them.

As one of the typical schemes of proactive routing, DSDV is applied to the basic routing protocol of multi-robot network systems. The principle of the scheme of DSDV is explained in the next subsection.

8.2.2 DSDV Routing Protocol

In DSDV, the routing table that each terminal constructs for routing processes is called *forwarding table* (from now on, denoted by *FT*). When a terminal intends to transmit a packet to some destination, this source terminal *forwards* the packet to one of neighboring terminals according to the current information of FT (this next terminal is called *next hop*). On the other hand, when a terminal receives a packet and it must be sent to some other destination, this relay terminal forwards the packet to the next hop as the source terminal did. This packet relaying repeats until the packet reachs the destination.

For the purpose to update the FT along with the network topology changes, each terminal periodically repeats the broadcasting of update packets and the reception of them. The broadcasting is done to all of the neighboring terminals and the reception is followed by the update of the own FT. This process of update packets exchanges and successive FT updates among neighboring terminals gradually floods the whole network with the current information of network configuration.

8.2.3 Forwarding Table and Routing Processes in DSDV

An example of ad hoc network and FT of each terminal based on DSDV routing protocol is shown in Figure 8.1. Four terminals A, B, C, and D configure the network in this example. A transmission range is defined for each terminal as a circle with predetermined transmission radius. Pair of terminals within the range of each other (or equally, closer than the transmission radius) establish a link (or, communication link). In Figure 8.1, the pairs of terminals A and B, B and C, and B and D establish the links. A pair of terminals, which locate out of the transmission range of each other, communicate through a path which traces a cascade of links. For example, terminal A and C communicate through the path A-B-C. If the source terminal is A, the relaying terminal B is determined as the next hop to the destination C according to the FT of A.

Examples of FT of each terminal at some instant during the ad hoc networking are indicated in Figure 8.1. In each FT, the following five terms are recorded in one line (the set of these terms is called *entry*):[1]

dst represents the destination.

nxt represents the next hop.

mtr represents the metric to the destination.

seq represents the sequence number of this entry.

ist represents the installed (or updated) time of this entry.

The pair of 'dst' and 'nxt' indicates the next hop to which the terminal forwards a packet that must be finally transmitted to 'dst'. For example, if the terminal A intends to transmit packets to C according to the FTs shown in Figure 8.1, the transmission process is as follows. First, A forwards the packets to B according to its FT that indicates the 'nxt' is B when 'dst' is C. Second, B receives the packets and forwards them to C according to its FT

[1] The descriptions in this subsection follow the ad hoc network simulator [300] constructed based on the original proposal of DSDV [296].

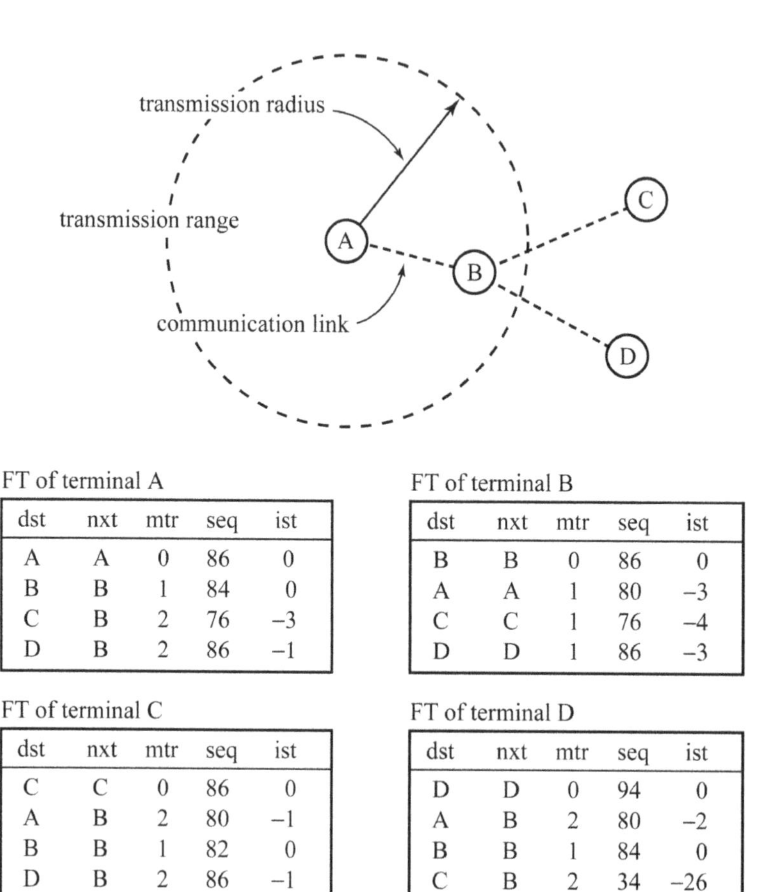

FT of terminal A

dst	nxt	mtr	seq	ist
A	A	0	86	0
B	B	1	84	0
C	B	2	76	–3
D	B	2	86	–1

FT of terminal B

dst	nxt	mtr	seq	ist
B	B	0	86	0
A	A	1	80	–3
C	C	1	76	–4
D	D	1	86	–3

FT of terminal C

dst	nxt	mtr	seq	ist
C	C	0	86	0
A	B	2	80	–1
B	B	1	82	0
D	B	2	86	–1

FT of terminal D

dst	nxt	mtr	seq	ist
D	D	0	94	0
A	B	2	80	–2
B	B	1	84	0
C	B	2	34	–26

Figure 8.1 Ad hoc network and FT of each terminal based on DSDV routing protocol

that indicates the ‘nxt’ is C when ‘dst’ is C. Finally, C receives the packets and the transmission process is accomplished.

On the other hand, ‘mtr’ and ‘seq’ are used to determine whether or not each entry is overwritten when an update packet is received from another terminal. ‘mtr’ means how far the location of ‘dst’ is and usually the hop count to the terminal indicates this value. As mentioned before, the hop count means how many links the path to ‘dst’ consists of. ‘seq’ indicates the oldness of the entry. Every entry is generated initially at ‘dst’ terminal and is broadcast periodically with a 'seq' number increased by two at each broadcasting.[2] For example, the 'dst' and 'seq' of the second entry in D's FT are A and 80 respectively. This indicates that this is 40th entry generated at terminal A.

When a terminal receives an entry as a part of update packets, it adds this entry to its FT provided the same 'dst' does not exist as the received one. However, if the entry already exists with the same 'dst' in its FT,

the original entry is overwritten with the received one provided the following two conditions are satisfied, or, the received one is discarded.

1. The 'seq' of the received entry exceeds that of the original one. This means that the received entry is generated later than the original one at the terminal 'dst' and therefore gives more recent information of the network topology.
2. If these 'seq's are equal to each other, the 'mtr' of the received entry falls behind that of the original one. This means that these entries are copies of one generated at 'dst' and the received one is transmitted through a shorter path than the other.

When the received entry overwrites the original one, 'nxt' is changed to the terminal that sent the update packet and 'mtr' is increased by one because the hop count to 'dst' increases one than that from the terminal of the new 'nxt'.

Finally, 'ist' means how many broadcasts of update packets are done by the terminal since it installed or updated the entry. If 'ist' of an entry exceeds a predetermined limit, and if the 'dst' of which establishes a link with the terminal, the terminal recognizes that the link is broken and does the following processes.

1. Add one to each 'seq' of every entry the 'nxt' of which is equal to the 'dst'.
2. Broadcast these entries with odd 'seq' as *warning* packets.
3. Erase these entries from its FT.

Every terminal that receives these entries with odd 'seq' rebroadcasts the warning packets and erases the corresponding entries from its FT.

When enough operation time is passed after a network topology change occurs, above processes of exchanging update packets or warning packets among terminals and successive FT updates of them in DSDV routing pro-

[2] This increment keeps the 'seq' number even indicating the availability of this entry. Whereas, if an update packet contains an odd 'seq', corresponding entry indicates the link breakage occurring as described later.

tocol, assure that the path exists between terminals and is directed by FTs of intermediate terminals provided every link is available to configure the path. Moreover, optimality of the path is assured, meaning that the hop count of the path is the minimum in comparison with other paths in the network.

8.3 Autonomous Chain Network Formations by Rescue Robots

In the operation of a multi-robot rescue system, its ad hoc network must form a chain configuration when distant spaces exist in disaster areas and the spaces must be reconnoitered by a finite number of robots. The distant spaces are within buildings or such structures and are distant from the safety zone where the BS is established. In this case, chain networks must be formed that stem from the BS and expand to the target spaces.

Regarding chain networks, the scenario of rescue operation is as follows. First, robots gather around the BS and only single hop communication links exist. Second, robots begin their collaborative movement to one of the distant target spaces and a chain network gradually emerges. Third, the chain reaches the target space and the robots in the forefront explore there. Fourth, the robots change the target of exploration to another space and the chain transforms along with their successive movements. Finally, the scenario returns to the third stage.

During this operation of chain networking, two points must be satisfied. First, because immediate consultation of the sensed data is important in rescue operations, the chain must always exist to assure the communications among robots and operators at BS. Second, for the purpose to avoid the traffic jamming with control signals that possibly occurs in chain networks, formations and transformations of chains must be executed by autonomous movements of robots. This autonomy of robots also avoids a regional system blackout. This blackout means that every robot within a region simultaneously ceases their entire functions. If an occasional breakage of chain network occurs and if robots must always be controlled by operators, this system trouble occurs beyond the breakpoint of the network.

In addition, robots must determine the exploration target of distant spaces without provided maps or such geometrical information about the disaster areas because the information may not be available in emergencies or become invalid after a destructive disaster.

In this section, a method is introduced for autonomous chain network formation by multi-robot rescue systems that reconnoiter distant spaces in disaster areas [301]. For the purpose to satisfy the above mentioned requirements, autonomous classification of robots into search robots (*SRs*) and relay robots (*RRs*) is adopted with behavior algorithms of each different classes of robots (this method is referred to as *search and relay robots classification* or *SRRC*). The rule of the classification and behavior algorithms refer to the forwarding table of each robot, constructed for ad hoc networking. SRs reconnoiter the distant spaces as forefronts of chain networks and RRs act as the intermediate nodes between the SRs and BS. Based on their collaborative movements, chain networks always exist and are formed autonomously according to the scenario of rescue operation.[3] For the purpose to explore unknown regions, SRs adopt decision-theoretic approach as an additional part to the behavior algorithm [303].

This section is organized as follows. In Section 8.3.1, the details of the proposed SRRC system procedure are explained. In Section 8.3.2, simulation results are presented in order to validate and confirm the suitability of the autonomous chain network formation executed by robots with proposed SRRC method.

8.3.1 Search and Relay Robot Classification

In SRRC, robots classify themselves autonomously into search and relay robots (*SRs* and *RRs*). SRs explore disaster area and RRs relay packets transmitted between SRs and the BS.

An SR is one which position is an endpoint of the network. When a robot recognizes that it is at this position, it classifies itself as an SR by recognizing the position based on a predetermined rule (called the *rule of SR/RR classification*). SRs explore the disaster area based on a predetermined behavior algorithm where exploration strategy is included.

An RR is one that recognizes itself as not being an SR. Each RR selects its master search robot (*MSR*) from the SRs in the network based on a predetermined rule (called the *rule of MSR selection*). An RR acts as a relay terminal within the path from its MSR to the BS and moves appropriately to maintain the path based on the behavior algorithm as SR does.

[3] An experiment of similar chain formation by real robots has been executed [302]. However, because of some different characteristics of this experiment exist with respect to SRRC, including fixed role of each robot in search and signal relaying, the performance of the scheme adopted in [302] may be restricted in actual disaster areas.

As mentioned above, there are two rules in SRRC processing: the rule of SR/RR classification and that of MSR selection. In addition, a behavior algorithm is specified, which each robot follows depending on its class.

In the following subsubsections, these rules and the behavior algorithm are described. System operation of multi-robot system is described afterwards including chain network formation and transformation executed by the system with SRRC.

8.3.1.1 Rules for SRRC Processing

The rules for SRRC processing are based on the forwarding table stored by each robot for the routing process of the ad hoc network. As mentioned in the previous section, DSDV is adopted as a basic protocol for ad hoc networking. The forwarding table and the rules for SRRC processing are explained using a simplified multi-robot system shown in Figure 8.2.

(1) *Forwarding table*

Figure 8.2 shows nine robots A, . . . , I and BS,[4] from which robot C is selected to show an example of the forwarding table in Figure 8.3. The table consists of entries for each destination (denoted by *dst*). Each entry records the elements associated with the destination and are updated repeatedly by the exchange of update packets with other robots in the network. As shown in this figure, elements are classified into inherent ones in DSDV and introduced ones for SRRC processing (*DSDV* and *SRRC elements*, respectively).

DSDV elements consist of *dst*, *nxt*, and *mtr* (other elements *seq* and *ist* are omitted here). In particular, the next hop recorded in the entry for BS is called *NBS*. The NBS of robot C is robot B, as shown in the first row.

The SRRC elements consist of *location*, *inv*, *cls*, and *msr*. *location* indicates where the destination is located in the global coordinate system.[5] *inv* indicates an inverted NBS that indicates which robot recognizes robot C as its NBS. As described later, SR/RR classification is based on *inv*. *cls* indicates the classification results of each destination of the robot. *msr* indicates the MSR of robot C. This *msr* is selected from SRs listed in the *cls* column.

Each robot transmits its own *location* and *cls* to all other robots by attaching them to update packets. Whereas, it transmits its NBS to

[4] Graphical notations indicating SR, RR, and BS are also used in the following figures.

[5] The localization of robots that explore known or unknown areas and their local map merging are fundamental problems and have been investigated so far [304–307]. However, in this paper we simply assume that global maps of robots are available with sufficient accuracy provided they can communicate with each other.

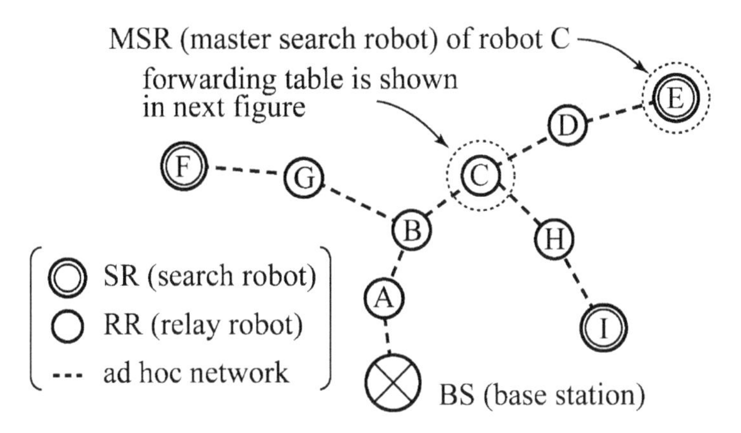

Figure 8.2 Simplified example of multi-robot system

NBS (next hop to BS)

dst	nxt	mtr	location	inv	cls	msr
BS	B	3	(223.5, 466.8)	–	–	
A	B	2	(210.9, 502.1)		RR	
B	B	1	(222.4, 533.6)		RR	
D	D	1	(293.6, 574.4)	*	RR	
E	D	2	(114.4, 568.4)		SR	*
F	B	3	(151.6, 564.4)		SR	
G	B	2	(189.6, 554.4)		RR	
H	H	1	(271.6, 519.6)	*	RR	
I	H	2	(288.4, 485.6)		SR	
← DSDV →			← SRRC →			

Figure 8.3 The forwarding table of robot C

neighboring robots only. This information makes *inv* of the neighboring robots. For example, robot D transmits that its NBS is robot C to neighboring robots C and E. Among them, robot C records this information in its *inv* because the NBS indicates robot C itself. Similarly, robot C records robot H in its *inv*. As a result, column *inv* shows D and H as an inverted NBS of robot C.

(2) *Rule of SR/RR classification*
As mentioned before, each robot classifies itself as SR when its position is the endpoint of the network, or as RR otherwise. This topological position of each robot is simply determined by *inv* column in its forwarding table.

If nothing is listed in the column, then the robot recognizes its position as an endpoint and classifies itself as SR. On the other hand, if one or more indexes are listed in *inv*, then the robot recognizes its position as not an endpoint and classifies itself as RR.

In Figure 8.2, because robots E, F, and I find no indexes in their *inv* column, they classify themselves as SRs. Whereas, because other robots have one or more indexes in their *inv*, such as robot C shown in Figure 8.3, they classify themselves as RRs.

Based on this rule of SR/RR classification, every endpoint of the network obviously becomes SR. However, some nodes may be classified as SRs even though they are not actual endpoints. These nodes are called *virtual endpoints* in SR/RR classification. Examples of virtual endpoints are shown later in Figure 8.5.

(3) *Rule of MSR selection*
Each RR selects its own MSR among the SRs in the network and moves appropriately to maintain the path between the MSR and BS. These robots must form a chain network where MSR and BS terminate both the endpoints and RR is at some intermediate position.

Satisfying this requirement, MSR of each RR must be a furthest robot from BS. This is because the furthest one possibly is the forefront of the chain network and RRs must support this SR by extending the chain as long as possible.

The rule of MSR selection concerning these requirements is as follows (the RR selecting its MSR is denoted by RR^s):

1. Among SRs indicated in the forwarding table as destinations, discard those for which the next hop coincides with NBS. This is because SR and BS beyond the same next hop configure some other chain which does not include RR^s.
2. Among the remaining SRs from the above screening, select the one having the largest distance from BS. This distance is measured along the optimal geographical path avoiding obstacles if they exist between each SR and BS.

For example, robot C in Figure 8.2 selects robot E as its MSR by the following selection process.

1. Robot C recognizes robots E, F, and I as SRs because they are indicated as SRs in the *cls* of its forwarding table shown in Figure 8.3.
2. Among these SRs, robot F is discarded based on the first item of the rule of MSR selection because the next hop to F and NBS are the same robot B.
3. Robot C calculates the distances from BS to robots E and I depending on their locations indicated in the forwarding table and the global map of the explored regions so far. Then, based on the second item of the rule of MSR selection, robot E is finally selected as the MSR of robot C.

8.3.1.2 Behavior Algorithm of Robots

After the SR/RR classification finishes, each robot acts according to the predetermined behavior algorithm to form a chain network to reconnoiter one of the distant spaces and transform it to reconnoiter another one. The details are as follows.

Each RR acts to maintain the link to the next hop to its MSR (denoted by *NMSR*). Simultaneously, it acts to maintain the link to its NBS. Such RR behavior depends on whether distance d_{out} to NMSR exceeds threshold r_{out} and whether distance d_{in} to NBS exceeds threshold r_{in}. r_{out} and r_{in} are called *outer link recovery radius* and *inner link recovery radius*, respectively. These values are fixed in relation to transmission radius r_t as $r_{in} < r_{out} < r_t$. Because the link to NBS is assumed to be more essential, r_{in} is set smaller than r_{out}. Distances d_{out} and d_{in} can be obtained from the power level of the signal received from NMSR and NBS, respectively.

On the other hand, each SR moves to its next target. This target is selected according to the decision-theoretic approach [303]. In this approach, the border of known and unknown regions is divided into cells and are called *frontier cells*. Each SR selects the next target among the frontier cells negotiating with other SRs. This selection is done at the beginning of system operation and when each SR arrives to the current target. Besides the movement to the next target, each SR simultaneously acts to keep the link to its NBS when d_{in} exceeds r_{in} as RR does.

The behavior algorithm that each robot follows is shown in Figure 8.4 as a flowchart. First, each robot determines its class based on the above rule of SR/RR classification. In the following, the behavior algorithm of each class is described.

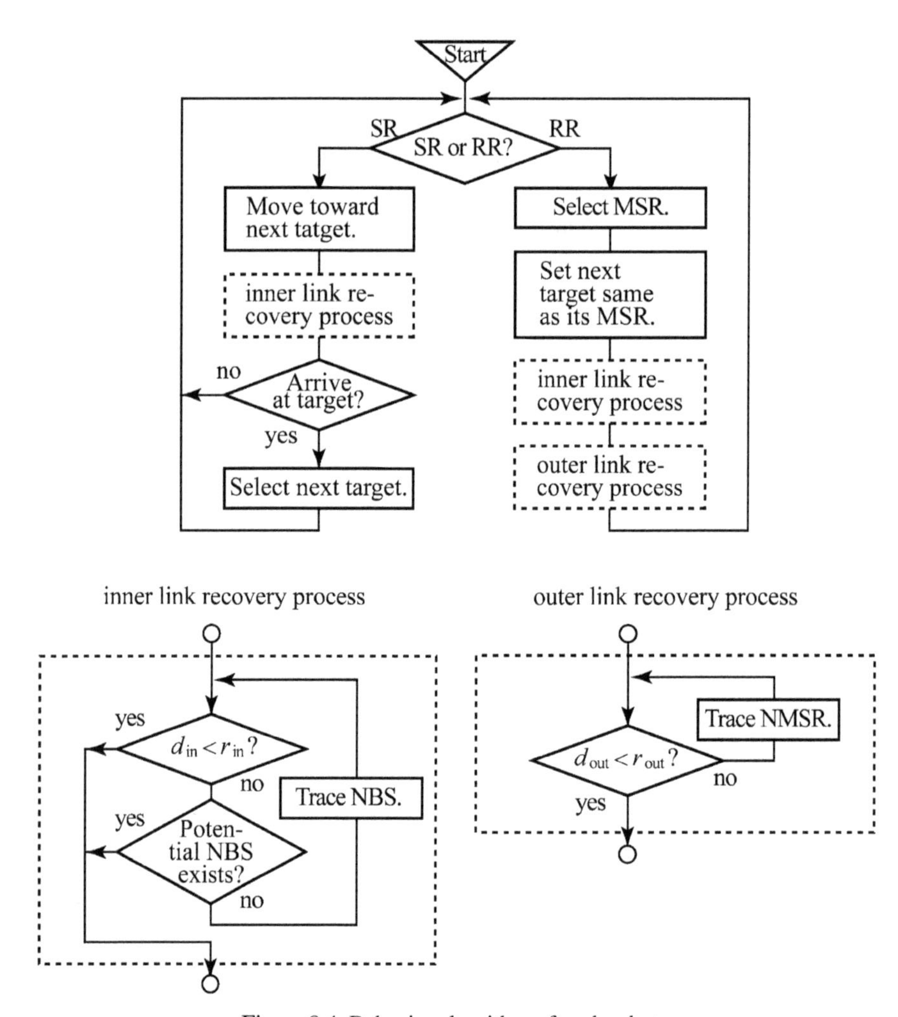

Figure 8.4 Behavior algorithm of each robot

(1) *Behavior algorithm of SR*
If the class is SR, then the robot moves toward the next target selected according to the decision-theoretic approach. Simultaneously, SR performs an *inner link recovery process* as follows.

If the robot finds that the distance d_{in} from its NBS exceed the threshold r_{in}, this means that the distance approaches the transmission radius r_t and will possibly exceed this critical value.

In this case, the robot examines whether or not another neighboring robot exists and the neighbor has path to BS. This neighboring robot is called *potential NBS*. If distance d_{in} to NBS exceeds the threshold r_{in} and a potential NBS is found, SR continues its movement. On the other hand, if no potential NBS is found, SR begins to trace[6] its current NBS until d_{in} becomes less than r_{in}.

Because the potential NBS is taken into account other than current one, the above mentioned process allows each SR proceed to the next target keeping the path to BS provided potential NBSs exist around the SR.

(2) *Behavior algorithm of RR*
On the other hand, if the class is RR, the robot selects its MSR according to the rule of MSR selection before mentioned. Then, it sets the next target to be the same as the one of its MSR. This target does not affect the behavior of RR. However, this memorized target becomes valid when the RR changes to SR. Because of this process, when a SR finishes the search of a space, the chain network with the SR can transform itself to reconnoiter another space. An example is shown later with Figure 8.5.

Besides these processes, the RR acts to maintain the path from the MSR to BS as follows:

1. If $d_{in} < r_{in}$ and $d_{out} < r_{out}$, then the robot recognizes itself as a relay terminal that is performing properly, so it stays there.
2. If $d_{in} \geq r_{in}$, then the robot decides that the link to the NBS is going to exceed r_t. In this case, it does inner link recovery process as mentioned before in the case of SR.
3. If $d_{out} \geq r_{out}$, then the robot decides that the link to the NMSR is going to exceed r_t. In this case, the robot simply traces the NMSR until d_{out} become less than r_{out}. This process is called *outer link recovery process*.

8.3.1.3 System Operation of Multi-Robot System

At the beginning of system operation, robots gather around the BS and perform ad hoc networking. When the network is completed, each SR determines its next target and moves to reach there. Because the farthest SR from the BS is selected as the MSR of many RRs, this SR and RRs gradually form a chain network threading a corridor to a distant space.

6 Here, to trace a robot means to approach the robot along the shortest path calculated taking into account the movement of the target.

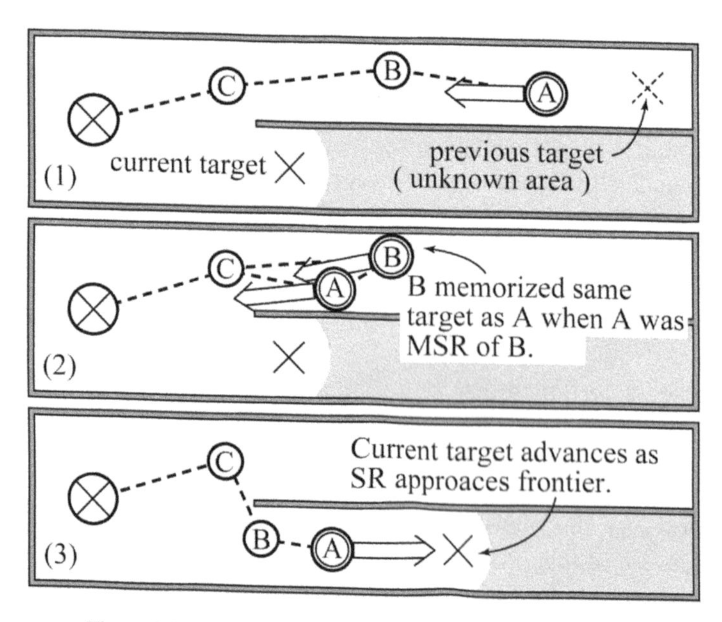

Figure 8.5 Autonomous transformation of a chain network

When the space has been searched, the SR updates the next target and the chain network transforms itself to reconnoiter another space beyond the new target.[7] This process of the chain network transformation is explained with Figure 8.5 where only two blind corridors exist for the simplicity.

The part (1) of this figure shows when the SR (robot A) reaches the end of a corridor and its next target is updated to one of frontier cells. This target locates in front of the unknown region of another corridor indicated by *current target*. At this moment, RRs (robot B and C) also update their target to the same point as their MSR (robot A). In the part (2), robot B becomes SR after some movement of A and, because B memorizes the same target as A, both robots move to the same next target. In the final part (3), because the SR (robot A) approaches the current target within frontier cells, the frontier advances along with the corridor and other robots follow A to extend the chain network.

[7] The same process will be triggered when the number of robots is not sufficient to reach the space, and therefore the SR cannot go ahead more for a predetermined time.

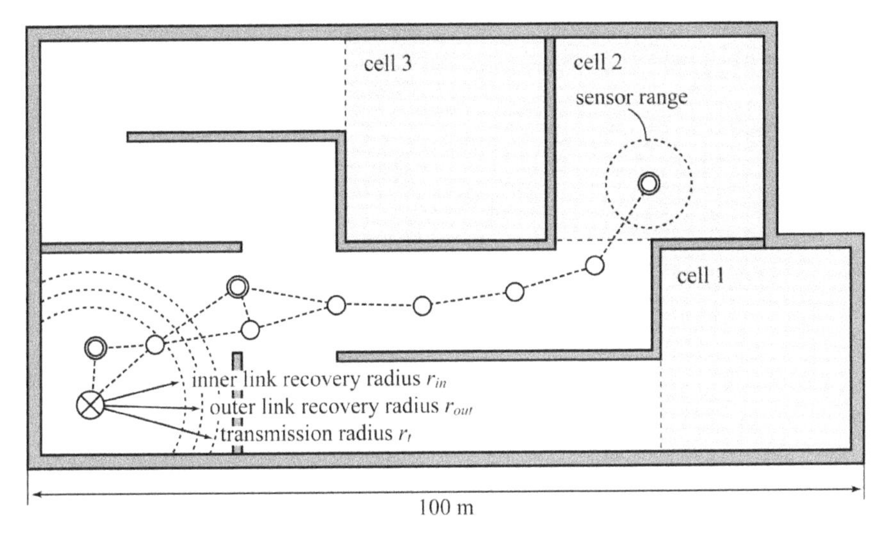

Figure 8.6 Simulation model of disaster area with distant spaces

8.3.2 Computer Simulation

The autonomous chain network formation executed by multi-robot system with the proposed SRRC system procedure is confirmed by computer simulations.[8] In this section, simulation model and simulation results are described.

8.3.2.1 Simulation Model

The simulation model is shown in Figure 8.6. Three distant spaces indicated by cells 1, 2, 3 exist beyond corridors stemming from a safety zone where BS is established.

The performance of chain network formation is measured here by the index of system operation: *search time*. This index indicates the elapsed time of the system operation when every cell is scanned entirely by robots. Here, a point in a cell is assumed to be scanned when the sensor range of a robot includes the point and simultaneously the robot has a path to the BS. The search time indicates how well the formation and transformation of chain networks are executed along with the system operation of reconnaissance to distant spaces.

[8] The simulator is constructed based on modules of the autonomous multi-robot simulator [308] and DSDV ad hoc network simulator [300].

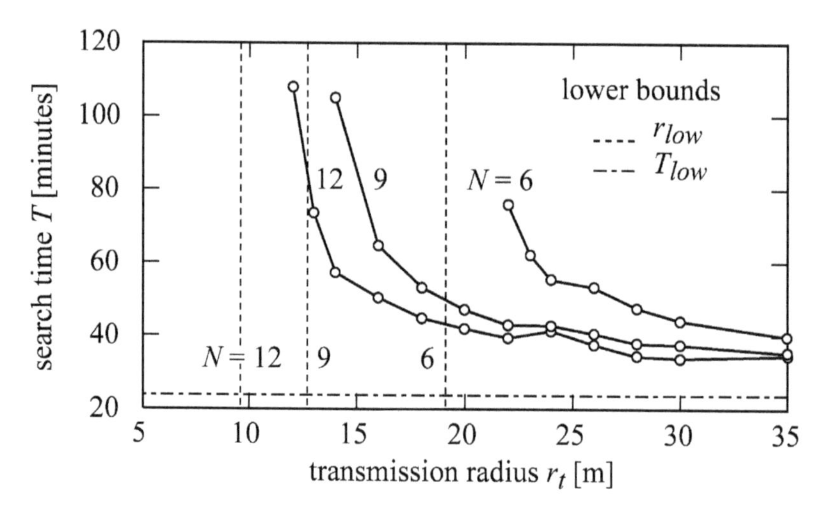

Figure 8.7 Search time over all cells

8.3.2.2 Simulation Results

Figure 8.7 shows the search time T over all cells 1, 2, 3 versus transmission radius r_t. The parameters are as follows. Outer link recovery radius r_{out} and inner link recovery radius r_{in} are set as $r_{out} = r_t - 2$ m, $r_{in} = r_{out} - 2$ m.

Within the distance r_t, update packets for FT updates are assumed to be transmitted always successfully. This means any collisions among packets do not occur on the assumption of the transmission rate of the packets being high enough. The frequency of forwarding table update is set to 10 times per second.

The sensor range is set to a circle with a radius of 5 m. The normal moving speed of a robot and the accelerated speed when it traces another one are set to 1 and 2 km/h, respectively.

There is a lower bound of r_t, denoted by r_{low}. If r_t is set to less than r_{low}, the exploration over the region becomes impossible. This bound is derived from the simple equation:

$$D = N \cdot r_{low} + r_s \tag{8.1}$$

where D is distance from BS to the farthest point within cells: the upper-right corner of cell 2, N is number of robots, and r_s is sensor range. Eq. (8.1) means that, provided $r_t > r_{low}$, the farthest point in the cells from BS can be within the sensor range of a robot at the endpoint of the longest chain. On the other hand, the time taken to reach the endpoint gives the lower bound of T such as:

$$D - r_s = s \cdot T_{low} \tag{8.2}$$

where T_{low} is the lower bound, s is the normal moving speed of robots. The lower bounds r_{low} and T_{low} are indicated by dashed and dashed-and-dotted lines, respectively.

Figure 8.7 shows that the transmission radius r_t possibly compensates the lack of number of robots N. When r_t exceeds about 30 m in Figure 8.7, the search time T approaches the lower bound T_{low} regardless of N. This is because few RRs are necessary to reconnoiter all cells, provided r_t is set large enough.

On the other hand, as r_t falls behind 30 m, T increases gradually and the performance difference by having different N appears. However, until r_t approaches the lower bound r_{low} of each N, it is confirmed that the reconnaissance into all cells is possible within reasonable time.

8.4 Wireless QoS Networks for Multi-Robot Systems

When a victim is detected by a robot during the operation of multi-robot rescue system, the robot must send the information to the BS with wideband signals such as dynamic pictures' images. Operators at BS observe these images and decide whether or not the information indicate a real victim. The wideband signals must be transmitted through the *wireless QoS network* where a specified signal bandwidth is reserved for every signals simultaneously transmitted from different robots to the BS. Furthermore, these signal transmission paths may suffer from unexpected breakages and successive reconstructions inherent in ad hoc networking.

In this section, a QoS routing method, the *rapid control method* or *RCM*, is introduced which is applicable to this operation of multi-robot systems where wideband signals must be transmitted to BS compromising with properties of ad hoc networking [309, 310]. The RCM utilizes a synchronized frame structure of time axis based on the DSDV routing protocol. Each path established in the network reserves a specified bandwidth and is called a *virtual circuit* or *VC*. When a VC is broken by a network topology change, the source terminal reestablishes the VC and transmits the residual data to the destination. This method is based on the *synchronized QoS routing* [311] with a modification of its frame structure that improves the *rate of successful communications*, especially when the network topology changes frequently.

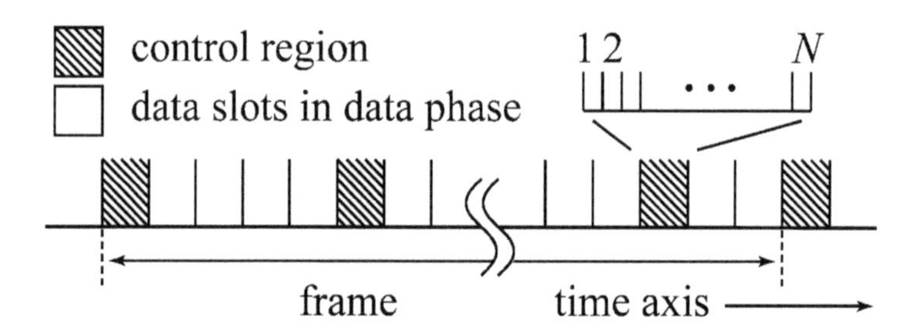

Figure 8.8 Frame structure of RCM (rapid control method)

This section is organized as follows. In Section 8.4.1, the details of the RCM are described in comparison with the synchronized QoS routing. In Section 8.4.2, to confirm the performance of RCM, results of computer simulations are shown adopting a special case of operation of multi-robot rescue system where wideband signals are transmitted to BS coincidentally by robots.

8.4.1 Rapid Control Method

In this subsection, the RCM is described focusing on its frame structure in comparison with the original structure in the synchronized QoS routing. Following this description, we explain why RCM improves the rate of successful communication. Finally, we outline the procedure of VC operation.

8.4.1.1 Proposed Frame Structure

Figure 8.8 shows the frame structure of RCM, which divides the time axis equally and which we assume to be synchronized over the network.[9] Each frame is subdivided into two phases: the control phase and the data phase. Control phase consists of *control regions* where every terminal has its own slot for control functions. In the case of N terminals forming the network, there are N slots in one control region as shown in the figure. The control function includes FT (forwarding table) updates and available bandwidth calculation. A restricted number of data slots is provided in the data phase for data packet transmission.

[9] This assumption may be satisfied if BS supplies every terminal with synch signal through the network. Or, decentralized scheme may be possible if MTSP (multi-hop time synchronization protocol) [312] is incorporated with RCM.

In this frame structure, multiple control regions are allocated and are scattered uniformly among data slots as shown in Figure 8.8. On the other hand, in the original frame structure of synchronized QoS routing, only one control region is allocated in each frame. Because this augmentation of control regions accelerates the transmission of control packets, obviously the call setup time and the transmission breakage interval both decrease. Furthermore, this acceleration of control processes rapidly removes idle VCs from the network, resulting in an improved rate of successful communications. This advantage of RCM is explained in the next subsection.

8.4.1.2 The Rate of Successful Communications

The rate of successful communications is defined as the average probability of a call setup and successive VC transmission both being accomplished. Here, the accomplishment of VC transmission means that the source terminal has transmitted the data to the destination for a predetermined time duration. VC breakages are allowed during this data transmission provided the VC is reestablished for the residual data transmission. The averaging of probabilities is done over all terminals and over all of their call trials during the system operation. This rate of successful communications is affected by idle VCs because of the following reason.

Each VC established in the network threads a cascade of communication links connecting neighboring terminals. When one of the links disappears, the VC breaks into two fragments. These fragments are called idle VCs. Although an idle VC cannot transmit any data, it occupies the bandwidth of relay terminals until control packets from the break point reach the terminals and release their idle bandwidth. Although each bandwidth occupation is temporary, idle VCs generally reduce the network capacity especially when they appear frequently due to changes in network topology. This reduction of network capacity tends to cause call setup process to fail, resulting in the deterioration of the rate of successful communications.

This degradation is alleviated by the frame structure of RCM because it accelerates the control packet transmission, which rapidly releases the bandwidth occupied by idle VCs. Therefore, the average of network capacity suffers less reduction than when the original frame structure is used. As a result, RCM improves the rate of successful communications.

8.4.1.3 Procedure of VC Operation

Figure 8.9 depicts an example procedure of VC operation where terminal A establishes a VC to the destination D and transmits data packets to there for

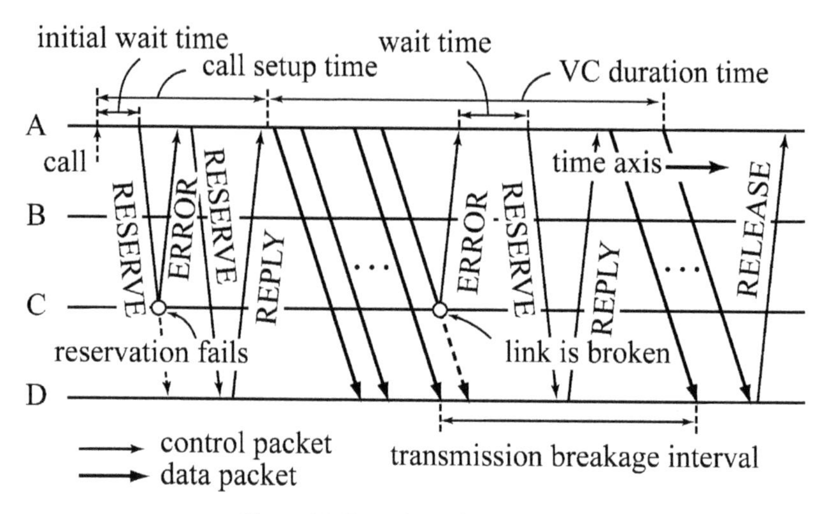

Figure 8.9 Procedure of VC operation

a predetermined duration. The VC is assumed to have two slots of bandwidth and is relayed by intermediate terminals B and C. This procedure operates as follows.

First, terminal A makes a call and determines the path to the destination D. In the case that the appropriate path is not found immediately with more than or equal to two slots of available bandwidth, possibly because other VCs are occupying the data slots of intermediate links, terminal A waits until the path appears, scanning its FT and *bandwidth table* repeatedly.[10] After this initial wait time has passed, terminal A transmits RESERVE to terminal D.

Second, terminal A receives a REPLY successfully from terminal D, or undesirably receives an ERROR from relay terminals B or C. In the former case, required data slots are reserved through the path for the intended VC. On the other hand, in the latter case, terminal A recognizes that the reservation has failed at the relay terminal. In this case, terminal A repeats the sending of RESERVE to terminal D until it receives a REPLY from the destination.

Third, because a REPLY has been received successfully, terminal A recognizes that a VC with the specified bandwidth is established and thus it begins the transmission of data packets to the destination D. The time interval between the initial call and the VC establishment is called the *call setup time*.

[10] Bandwidth table indicates available bandwidth of each path to every destination recorded in the FT. Construction of this table is a fundamental of *synchronized QoS routing* [311].

Finally, terminal A accomplishes the data packet transmission, or, before the final packet arrives at the destination, the source terminal receives an ERROR from one of the relay terminals. In the former case, the destination D sends a RELEASE to the source terminal. This control packet releases all the data slots reserved throughout the VC. On the other hand, in the latter case, terminal A recognizes that a link is broken at the relay terminal. In this case, terminal A repeats the process of call setup for the recovery VC. Until the residual data reach the destination D, the *transmission breakage interval* appears as shown in Figure 8.9.

8.4.2 Computer Simulations

Computer simulations are executed to confirm the system performance of multi-robot rescue system with RCM. A simple operation scheme of reconnaissance into distant space by the rescue system [313] is adopted as the simulation model. As described in the previous section, distant spaces are within buildings or such structures and are distant from the safety zone where the BS is established. This scheme of reconnaissance into the space consists of sequential movement of robots to a destination within the space, and random walking of robots after they reach the destination. A DSDV ad hoc network simulator [300] is used with modifications including a synchronized frame structure and VC operation based on the bandwidth calculation.

8.4.2.1 Simulation System Model

Figure 8.10 shows the simulation system model. In this model, two spaces are connected by a crooked path. The BS and the destination of robots are located at the center of the respective spaces. Five robots have already arrived at the destination and changed their behaviors to randomly walking around the destination. Eleven other robots are commanded to stop and remain stationary when the fifth robot begins to walk around. Consequently, there are now five walking robots around the destination, occasionally transmitting information to the BS, and eleven stationary robots that only relay the transmission through the network based on the synchronized QoS routing. Because only the robots that make calls are concerned, in the following discussion, the term "robot" means one of the five walking robots.

Other assumptions are as follows. Real-time streaming of MPEG-4 video signals is transmitted from robots to the BS. Signal profiles are at level 3 or 4 of ARTS (advanced real-time simple) [314]. Each robot ordinarily transmits a level-3 signal, and if the scene indeed shows a victim, a high-resolution

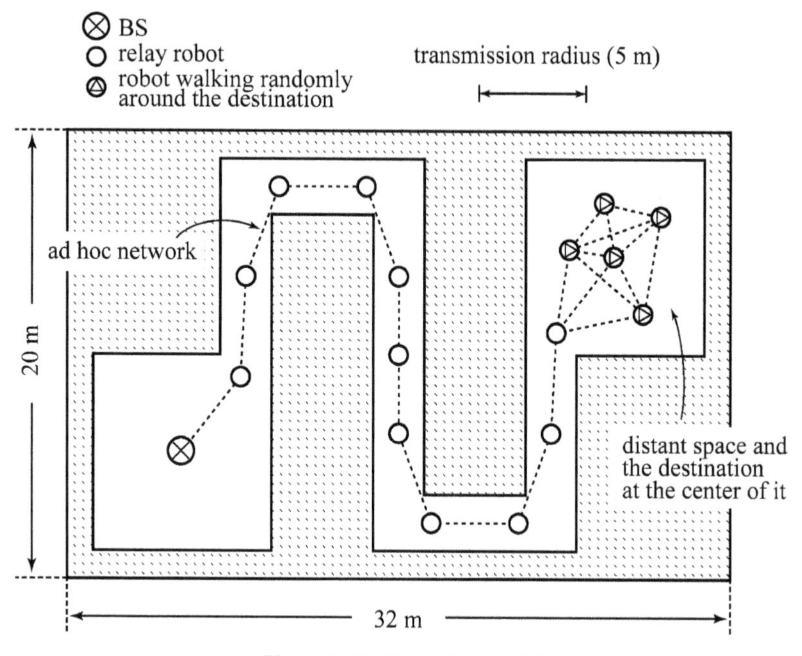

Figure 8.10 Simulation model

signal of level 4 is transmitted at the request of operators. Therefore, the call rate of each robot is defined by a pair of values (α, β), where α is for the level-3 signal and β is for the level-4 signal. Network traffic is classified here into three categories: heavy, moderate and light. For these cases, the pair of call rates (α, β) is set at (6, 0.6), (2, 0.2) and (0.6, 0.06), respectively. The call rate is defined here as the averaged call time per minute for each robot.

For the purpose of adjusting the frame structure to the MPEG-4 signal transmission, the transmission rate, the frame length, and the data slot length are set at 7.4 Mbits/sec, 92 msec, and 5 msec, respectively. Because one data slot transmits 402 kbits/sec, ARTS level-3 signals (384 kbits/sec) and level-4 signal (2 Mbits/sec) can be transmitted with the bandwidth of one data slot and five data slots, respectively. Each transmission is intended to continue for 30 seconds. The length of the control region is set at 2 msec divided equally into that of 17 control slots, each for every robot and BS in the network.

One control region and 18 data slots are included in one frame in the case of the original frame structure. This structure is indicated by (1, 18). On the other hand, three pairs of numbers, (6, 16), (11, 14), and (16, 12), are assumed in the case of the frame structure of RCM. These four structures result in the

transmission rate	7.4 Mbits/sec
frame length	92 msec
duration	30 seconds
network traffic: MPEG-4 (call rate of ARTS L3, L4) [time/min/terminal]	heavy (6, 0.6) moderate (2, 0.2) light (0.6, 0.06)
frame structure (number of control regions, number of data slots)	original (1, 18) proposed (6, 16) (11, 14), (16, 12)
call setup limit τ [seconds]	original 2.8 proposed 0.74

Figure 8.11 Simulation parameters

transmission rate equal to each other. These parameters are summarized in the table in Figure 8.11.

8.4.2.2 Simulation Results

In the following, simulation results are shown for the call setup limit, the rate of successful communication, the call setup time, and the transmission breakage interval. At each simulation, system operation is assumed to continue for 30 minutes.

(1) *Call setup limit*

As described before (see Figure 8.9), call setup time is necessary before each robot establishes the VC. When this time exceeds a threshold, i.e. the so-called *call setup limit* τ, the robot determines that the call trial has failed. τ should be set small to alleviate network congestion caused by stale signals. However, too small τ may prevent call setup processes being accomplished as depicted in Figure 8.9.

From the simulation result, we set τ to 2.8 and 0.74 seconds for the original (1, 18) and frame structure (6, 16) of RCM, respectively. For other RCM frame structures, 0.74 seconds is used similarly.

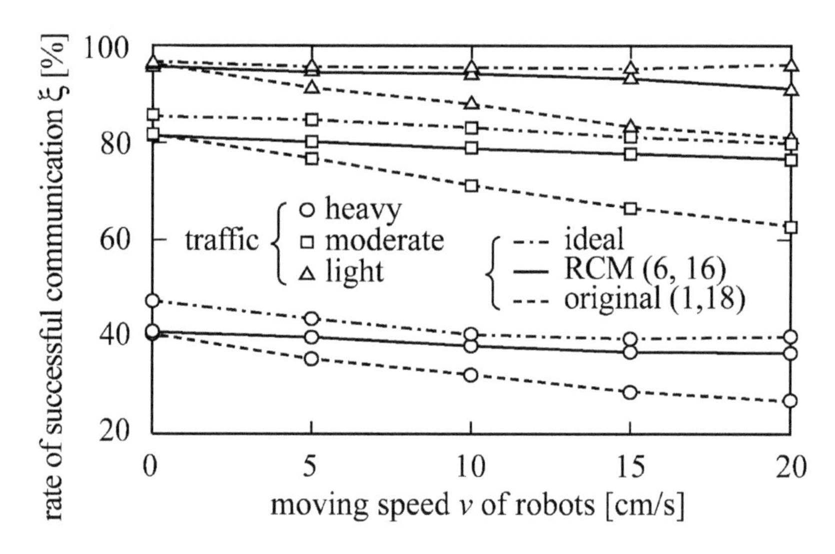

Figure 8.12 Rate of successful communications vs. moving speed of robots

(2) *Rate of successful communication*
Figure 8.12 shows the rate of successful communication ξ versus moving speed v of robots. The simulation result of the original frame structure (1, 18) and that of RCM (6, 16) are indicated by dashed and solid lines, respectively. Besides these structures, the ultimate case is also estimated where control packets are assumed to be transmitted instantaneously, whereas one frame includes 18 data slots as per the original structure. This ideal case is indicated by dashed-and-dotted lines.

As v increases, ξ decreases monotonically in every case because a large v causes frequent network topology changes resulting in a high possibility of VC breakage occurring. However, the frame structure (6, 16) alleviates this degradation of ξ by rapidly removing idle VCs from the network. For example, when v equals 20 cm/s and traffic is moderate, the RCM frame structure improves ξ by about 17% over the case of the original structure. This amount of improvement does not reach the ideal case being two data slots in one frame not sufficient.

As shown in Figure 8.12, the expansion of the control region improves the rate of successful communication. However, accompanying decrease of data slots affects availability of bandwidth to be reserved and therefore impairs the rate of successful communication. Figure 8.13 shows this tradeoff in the frame structure when robots are moving at 20 cm/s. When the number

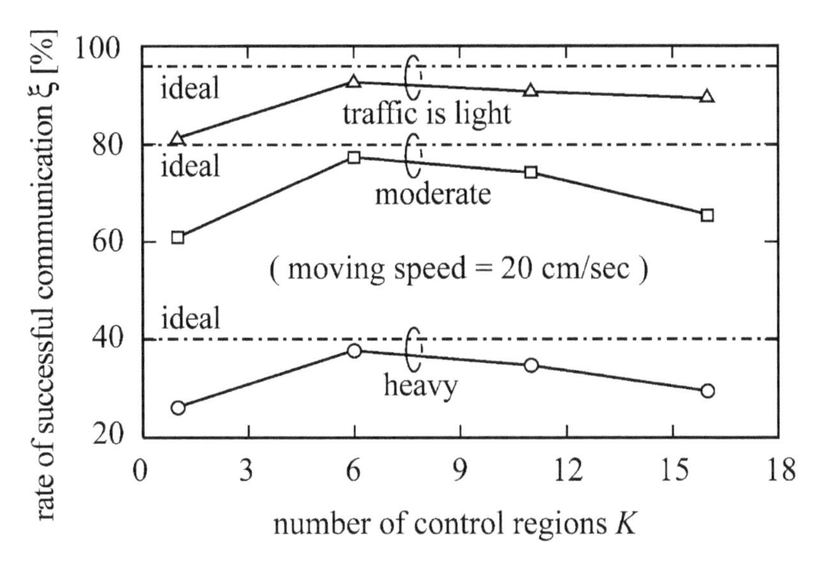

Figure 8.13 Rate of successful communication vs. number of control regions

of control regions K equals 1, this indicates the original frame structure with 18 data slots. Other cases indicate the RCM frame structures of (6, 16), (11, 14), and (16, 12). As this figure shows, the frame structure (6, 16) leads to the closest ξ with respect to the ideal case, over other structures and at any traffic volume. For this reason, the structure (6, 16) is adopted in the following investigations.

(3) *Call setup time and transmission breakage interval*

Figure 8.14 indicates the call setup time γ_s and the transmission breakage interval γ_i versus the moving speed v of robots. Here, γ_s is the average value over all the established VCs at every robot and is indicated by a dashed-and-dotted line. Furthermore, γ_i is the average value over the VCs where only the accomplished ones, in spite of breakages occurring, are taken into account. The γ_i of one breakage is indicated by a solid line and the summation of that throughout the VC is indicated by a dashed line.

As this figure shows, γ_s remains almost constant in both cases of the frame structure regardless of v. This is because the movement of robots scarcely affects the reservation process of initial VC setup. These constant values are 2.4 and 0.4 seconds in the case of the original and the RCM frame structures, respectively. Therefore, it is confirmed that unreasonable call setup

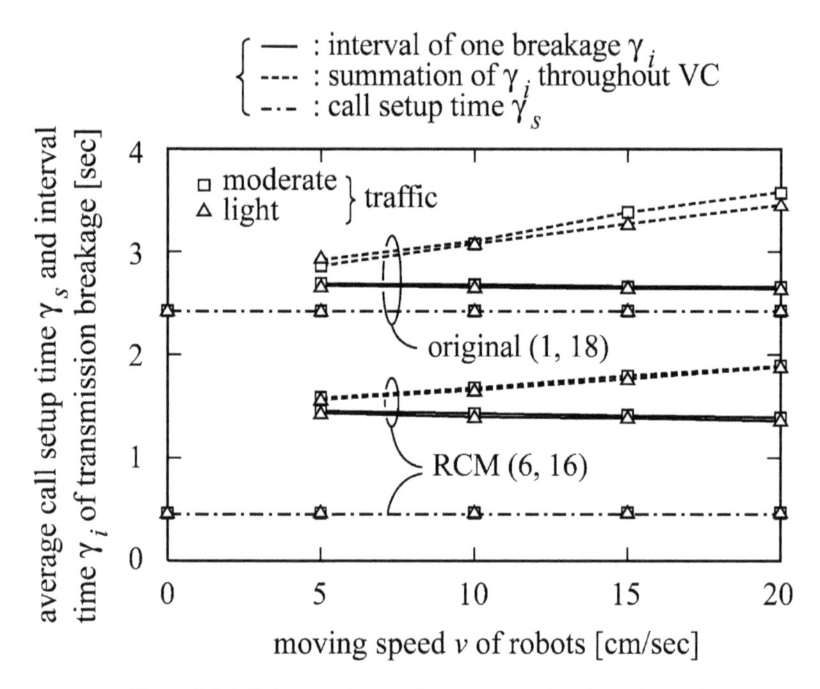

Figure 8.14 Call setup time and transmission breakage interval

times arise in the case of the original frame structure (1, 18), where each robot cannot transmit instantaneously the first scene of the detected victims, and that this problem is alleviated significantly by using the RCM frame structure (6, 16).

The simulation results also confirm that the transmission breakage interval γ_i becomes unreasonable when the original frame structure is used, and that this interval time decreases by about half when using the RCM frame structure at any v. However, the transmission breakage interval still assumes a considerable value even when the RCM scheme is used. Therefore, in this case, operators at the BS may still be disturbed sometimes when observing continuous scenes of victims. Some other schemes should thus be investigated to improve the efficiency of the operators' task that prevents interruptions due to transmission breakages.

8.5 Conclusion

Operations of multi-robot rescue systems are introduced focusing on their ad hoc networking established on wireless links of neighboring robots. Because of the advantages of ad hoc network including reliability and scalability as described in Section 8.2, this network scheme is suitable for multi-robot rescue systems performing their tasks in hazardous environment and communicating with operators at BS.

Two operation schemes of multi-robot rescue systems: autonomous chain network formations, and wireless quality of service (QoS) networking are introduced adopting DSDV routing protocol as a basic one in the proactive class for ad hoc networks.

The autonomous chain network formations by rescue robots are introduced in Section 8.3. For the purpose to assure the chains always exist and the formation is executed by autonomous movements of robots, SRRC (search and relay robots classification) is adopted as the operation scheme of multi-robot system. In SRRC, each robot classifies itself autonomously into SR (search robot) or RR (relay robot). Each SR mainly moves to its next target determined by decision-thoretic approach and each RR acts as a relay terminal within the chain network from its MSR (master SR) to the BS (base station). Computer simulations confirm the satisfactory performance of the chain network formations and transformations by rescue robots adopting SRRC, along with a specific model of disaster area.

The wireless quality of service (QoS) networks for multi-robot systems are introduced in Section 8.4. With the purpose to reserve transmission bandwidth between each robot and BS in ad hoc networking, RCM (rapid control method) is adopted based on synchronized frame structures over the network. RCM is a modification of the original scheme of synchronized QoS routing and improves the network performance over the original scheme, in terms of rate of successful communication, call setup time and transmission breakage interval. These improvements are confirmed by computer simulations where a simple operation scheme of reconnaissance into distant space by robots is considered.

Bibliography

[1] A.M. Townsend and M.L. Moss, Telecommunications Infrastructure in Disasters: Preparing Cities for Crisis Communications, Center for Catastrophe Preparedness and Response & Robert F. Wagner Graduate School of Public Service, New York University, http://www.nyu.edu/ccpr/pubs/NYU-DisasterCommunications1-Final.pdf, April 2005.

[2] S. Arroyo Barrantes, M. Rodriguez, and R. Prez (Eds.), *Information Management and Communication in Emergencies and Disasters – Manual for Disaster Response Teams*, Pan American Health Organization, Washington, DC, 2009.

[3] ICT for Disaster Management/ICT for Disaster Prevention, Mitigation and Preparedness, Wikibooks, http://en.wikibooks.org/wiki/ICT_for_Disaster_Management/ICT_for_Disaster_Prevention,_Mitigation_and_Preparedness.

[4] International Telecommunication Union, Handbook on Emergency Telecommunications, http://www.itu.int/ITU-D/emergencytelecoms/doc/handbook/pdf/Emergency_Telecom-e-chapter1.pdf, 2005.

[5] Enhancing Regional Cooperation on Disaster Risk Reduction in Asia and the Pacific: Information, Communications and Space Technologies for Disaster Risk Reduction, United Nations Economic and Social Council, Economic and Social Commission for Asia and the Pacific, Committee on Disaster Risk Reduction, http://www.unescap.org/idd/events/cdrr-2009/CDR_5E.pdf, March 2009.

[6] S. Szygenda and M.A. Thornton, Disaster Tolerant Computer and Communication Systems, Orlando, FL, 2005.

[7] M.A. Harper, C. Lawler, and M.A. Thornton, IT Application Downtime, Executive Visibility and Disaster Tolerant Computing, 2005.

[8] Defense Acquisition University, Systems Engineering Fundamentals, Defense Acquisition University Press, 2001.

[9] F. Lipson, A. Longstaff, and R. Mead, Survivability: Protecting your Critical Systems, *IEEE Internet Computing*, December 1999.

[10] B.M. Thursisingham and J.A. Maurer, Information Survivability for Evolvable and Adaptable Real-Time Command and Control Systems, *IEEE Transactions on Knowledge and Data*, pp. 228–238, January 1999.

[11] R.J. Ellison, R.C. Linger, T. Longstaff, and N.R. Mead, Survivable Network Systems Analysis: A Case Study, *IEEE Software*, pp. 70–77, July 1999.

[12] I. Koren and M. Krishna, *Fault Tolerant Systems*. Morgan Kaufmann Publishers, New York, 2007.

[13] Ministry of Defense, Requirements for Safety Related Software in Defense Equipment, 00-55 (Part 2)/Issue 2, ed. 1997.

[14] L.M. Shooman, *Reliability of Computer Systems and Networks. Fault Tolerance, Analysis, and Design*. Wiley-Interscience, 2002.

[15] T.F. Arnold, The Concept of Coverage and Its Effect on the Reliability Model of Repairable Systems, C-22, pp. 251–254, ed. 1973.
[16] J.-C. Laprie, Dependability of Computer Systems: From Concepts to Limits, Toulouse, France, pp. 2–11, 1995.
[17] F. Lipson and A.D. Fisher, Survivability – A New Technical and Business Perspective on Security, Association for Computing Machinery, 1999.
[18] Committee on the FORCEnet Implementation Strategy, Naval Studies Board, FORCEnet Implementation Strategy, The National Academies Press, 2005.
[19] Government Accountability Office, Critical Infrastructure Protection, Federal Efforts Require a More Coordinated and Comprehensive Approach for Protecting Information Systems, GAO-02-474, July 2002.
[20] M.G. Christel and K.C. Kang, Issues in Requirements Elicitation, Software Engineering Institute, Pittsburgh, Pennsylvania, CMU/SEI-92-TR-12, September 1992.
[21] R.R. Young, *Effective Requirements Practices*, 3rd ed. Addison-Wesley, 2001.
[22] P. Parviained, H. Hulkko, J. Kaariainen, J. Takalo, and M. Tihinen, *Requirements Engineering Inventory of Technologies*. VTT Publications, 2003.
[23] J. Karlsson, S. Olsson, and K. Ryan, Improved Practical Support for Large-Scale Requirements Prioitising, *Requirements Engineering*, vol. 2, pp. 51–60, 1997.
[24] Institute of Electrical and Electronics Engineers, IEEE Guide for Developing System Requirements Specifications, IEEE Standard 1223 ed. The Institute of Electrical and Electronics Engineers, New York, 1998.
[25] Assistant Secretary of Navy for Research, Development, & Acquisition, Naval Systems of Systems Systems Engineering Guidebook, II ed., 2006.
[26] J.M. Carlson and J. Doyle, Highly Optimized Tolerance: A Mechanism for Power Laws in Designed Systems, *Physcial Review*, vol. 60, pp. 1412–1427, April 1999.
[27] V. Rykov, B. Dimitrov, D.J. Green, and P. Stanchev, Reliability of Complex Hierarchical Systems with Fault Tolerance Units, Santa Fe, NM, 2004.
[28] J. Klion, A Redundancy Notebook, Rome Air Development Center; Air Force Systems Command, Rome, New York, December 1977.
[29] W.-K. Chen, *The Electrical Engineering Handbook*. Academic Press, 2005.
[30] S. Habinc, Functional Triple modular Redundancy (FTMR). VHDL Design Methodology for Redundancy in Combinatorial and Sequential Logic, Gaisler Research, FPGA-003-01, December 2002.
[31] R.E. Lyons and W. Vanderkulk, The Use of Triple-Modular Redundancy to Improve Computer Reliability, *IBM Journal of Research and Development*, vol. 6, no. 2, p. 200, April 1962.
[32] J.C. Tryon, *Quadded Logic, Redundancy Techniques for Computing Systems*. Spartan, Washington, DC, 1962.
[33] R.W. Prew, Integrated but Separate; Advances in Integrated Safety Control, Special Report Automation Systems, 2008.
[34] H. B. Jr. Blanton, Reliability of Series, Parallel and Quad Redundancies with Dependently Failing Components, Defense Technical Information Center, AD0425718, August 1963.
[35] D.P. Siewiorek and E.J. McCluskey, Switch Complexity in Systems with Hybrid Redundancy, *IEEE Transactions on Computers*, vol. C-22, no. 3, pp. 276–282, March 1973.

[36] M. Harper, Systems Engineering for the Development of Disaster Tolerant Systems, RMS Partnership, 2008.

[37] E.R. Berlekamp, Bit-Serial Reed-Solomon Encoders, *IEEE Transactions on Information Theory*, vol. IT-28, no. 6, pp. 869–874, November 1982.

[38] J.-L. Dornstetter, Method of Transmission, with the Possibility of Correcting Bursts of Errors of Information MEssages and Encoding and Decoding Devices for Implementing This Method, June 28, 1988.

[39] I.K. Proudler, Non-Separate Arithmetic Codes, Defense Technical Information Center, ADA202064, September 1988.

[40] H.H. Kollmeier, Reconfiguration for Fault Tolerance and Performance Analysis, University of Pennsylvania Department of Computer and Information Science, MS-CIS-87-106, 1987.

[41] E.M. Clarke and C.N. Nikolaou, Distributed Reconfiguration Strategies for Fault-Tolerant Multiprocessor Systems, C-31, pp. 771–784, ed. 1982.

[42] M.A. Harper and J. Stracener, Systems Engineering Analysis to Improve Concept Development of Complex Defense Systems, San Diego, CA, 2007.

[43] C. Lawler, M.A. Harper, and M.A. Thornton, Components and Analysis of Disaster Tolerant Computing, New Orleans, LA, 2007.

[44] T.M. Todinov, Reliability Analysis Based on the Losses from Failure, *Risk Analysis*, vol. 26, no. 2, pp. 311–335, 2006.

[45] R. Walpole, R. Myers, and S. Myers, *Probability and Statistics*, 6th ed. Prentice Hall, p. 35, 1998.

[46] Committee on Pre-Milestone A Systems Engineering, National Research Council, Pre-Milestone A and Early-Phase Systems Engineering: A Retrospective Review and Benefits for Future Air Force Acquisition. The National Academies Press, 2008.

[47] Department of Defense, DODI 5000.02 – Operation of the Defense Acquisition System, 2009.

[48] P. Young and S. Tippins, *Managing Business Risk*. American Management Association, New York, NY, 2001.

[49] M.A. Thornton, S.A. Szygenda, and S. Nair, Large Systems Design and the Aximatic Process, 2009.

[50] W. Wang, J. Loman, and P. Vassiliou, Reliability Importance of Components in a Complex System, 2004.

[51] International Telecommunications Union, Telecommunications Save Lives, http://www.itu.int/ITU-D/emergencytelecoms/doc/brochure2.pdf [Accessed December 26, 2009], October 2005.

[52] United Nations Office at Geneva, Tampere Convention: Saving Lives through Emergency Telecommunications, http://www.unog.ch/80257631003154D9/(httpNewsByYear_en)/80257631003154D9C12570F1004B60A0?OpenDocument [Accessed December 26, 2009], 7 January 2005.

[53] G.M. Grant, et al., Emergency Diesel Engine Generator Power System Reliability 1987–1993, Idaho National Engineering Laboratory, Idaho, US, INEL-9510035, February 1996.

[54] D.R. Kuhn, Sources of Failure in the Public Switched Telephone Network, *Computer*, vol. 30, pp. 31–36, April 1997.

[55] K. Yotsumoto, S. Muroyama, S. Matsumura, and H. Watanabe, Design for a highly efficient distributed power supply system based on reliability analysis, in *Rec. INTELEC 1988*, pp. 545–550, 1988.

[56] A. Kwasinski and P.T. Krein, Multiple-Input DC-DC Converters to Enhance Local Availability in Grids Using Distributed Generation Resources, in *Rec. APEC 2007*, pp. 1657–1663, 2007.

[57] V. Purani, APC, White Paper #69 – Power and Cooling for VoIP and IP Telephony Applications, http://www.apcmedia.com/salestools/SADE-5TNRLR_R0_EN.pdf, 2003.

[58] A. Villemeur, *Reliability, Availability, Maintainability, and Safety Assessment. Volume 1, Methods and Techniques*. John Wiley and Sons, West Sussex, UK, 1992.

[59] A. Towsend and M.L. Moss, Telecommunications Infrastructure in Disasters: Preparing Cities for Crisis Communications, Center for Catastrophe Preparedness and Response, New York University, April 2005.

[60] J.W. Simmons, Digging out after Hugo, in *Rec. INTELEC 1990*, pp. 8–15, 1990.

[61] J. Coppinger, Wind storms in Britain, in *Rec. INTELEC 1990*, pp. 2–7, 1990.

[62] R. Ireland, Learnings and Challenges of the Northern California Earthquake, in *Rec. INTELEC 1990*, p. 1, 1990.

[63] A. Kwasinski, Telecommunications Outside Plant Power Infrastructure: Past Performance and Technological Alternatives for Improved Resilience to Hurricanes, in *Rec. INTELEC 2009*, Inchon, South Korea, 6 pp., October 2009.

[64] A.J. Schiff (Ed.), Northridge Earthquake: Lifeline Performance and Post-Earthquake Response, ASCE Technical Council on Lifeline Earthquake Engineering, Monograph No. 8, August 1995.

[65] A.J. Schiff (Ed.), Hyogoken-Nanbu (Kobe) Earthquake of January 17, 1995. Lifeline Performance, ASCE Technical Council on Lifeline Earthquake Engineering, Monograph No. 14, September 1998.

[66] A.K. Tang (Ed.), Izmit (Kocaeli), Turkey, Earthquake of August 17, 1999; Including Duzce Earthquake of November 12, 1999. Lifeline Performance, ASCE Technical Council on Lifeline Earthquake Engineering, Monograph No. 17, March 2000.

[67] A.J. Schiff and A.K. Tang (Eds.), Chi-Chi, Taiwan, Earthquake, of September 21, 1999. Lifeline Performance, ASCE Technical Council on Lifeline Earthquake Engineering, Monograph No. 18, July 2000.

[68] M. Yashinsky (Ed.), San Simeon earthquake on December 22, 2003 and Denali, Alaska, Earthquake of November 3, 2002, ASCE Technical Council on Lifeline Earthquake Engineering, Monograph No. 28, April 2004.

[69] L.V. Lund and C. Sepponen (Eds.), Lifeline Performance of El Salvador Earthquakes of January 13 and February 13, 2001, ASCE Technical Council on Lifeline Earthquake Engineering, Monograph No. 24, September 2002.

[70] C. Strand and J. Masek (Eds.), Sumatra-Andaman Islands Earthquake and Tsunami of December 26, 2004. Lifeline Performance, ASCE Technical Council on Lifeline Earthquake Engineering, Monograph No. 30, August 2007.

[71] Hawaii Governor's Comprehensive Communications Review Committee, Recommendations for Improving Public Communications during Emergencies, Final Report, January 5, 2007.

[72] Per, Instituto Nacional de Defensa Civil (INDECI), Lecciones aprendidas del sur: Sismo de Pisco, 15 agosto 2007. Document No. 17541, Lima, INDECI, 234 pp., 2009.

[73] H.K. Miyamoto, Sichuan China Earthquake, May 12, 2008, Field Investigation Report, May 2008.
[74] A. Tang, Telecommunications Performance – May 2008, Wenchuan, Sichuan Earthquake, in *Proc. TCLEE 2009: Lifeline Earthquake Engineering in a Multihazard Environment*, pp. 1407–1415, 2009.
[75] J.K. Burton, Hurricane Isabel: Government-Industry Partnership in Planning, Response and Recovery, http://www.nric.org/meetings/docs/meeting_20031205/Isabel_for_NRICvLS_Dec05.ppt [Accessed December 27, 2009].
[76] The Florida State Emergency Response Team, Situation Report No. 15 Tropical Depression Dennis, July 13, 2005.
[77] Cameron Communications, Rita Couldn't Break the Connection, http://eliztel.com/pdfs/Hurricane_Rita_Brochure.pdf [Accessed December 27, 2009].
[78] Texas Public Utility Commission, Project No 32182 PUC Investigation of Methods to Improve Electric and Telecommunications Infrastructure to Minimize Long Term Outages and Restoration Costs Associated with Gulf Coast Hurricanes, Final Staff Report, August 11, 2006.
[79] Texas Public Utility Commission, Public Records.
[80] R.U. Silver, National Communications System (NCS). Local Loop Overview, http://www.ncs.gov/tpos/esf/homestead/Silver-Local%20Loop.ppt [Accessed December 27, 2009].
[81] A. Kwasinski, W. Weaver, P. Chapman, and P.T. Krein, Telecommunications Power Plant Damage Assessment Caused by Hurricane Katrina – Site Survey and Follow-up Results, *IEEE Systems Journal*, vol. 3, no. 3, pp. 277–287, September 2009.
[82] G. Bicket and M. Latino, Rebuilding Broadband Infrastructure Following Katrina – Lessons Learned, presented at Katrina Forum, New Orleans, Louisiana, USA, September 24–26, 2006.
[83] P.J. Aduskevicz, AT&T Response Terrorist Attack September 11, 2001, Presentation to NRIC V, http://www.authorstream.com/Presentation/Savina-15089-appendixd-ResponseTerrorist-AttackSeptember-11-2001-Summary-Key-Responses-Impact-September-11th-Terror-the-present-and-future-of-fare-system-standards-paul-ppt-powerpoint/ [Accessed December 27, 2009], October 30, 2001.
[84] V. Vittore, California Fires Prompt Telco Aid, Concern, in *Telephony Online*, October 29, 2003.
[85] A. Kwasinski, Technology Planning for Electric Power Supply in Critical Events Considering a Bulk Grid, Backup Power Plants, and Micro-grids. *IEEE Systems Journal*, vol. 4, no. 2, pp. 167–178, June 2010.
[86] A. Kwasinski, Power Electronic Interfaces for Ultra-Available DC Micro-grids, Presented at the 2nd IEEE International Symposium on Power Electronics for Distributed Generation Systems (PEDG2010), Hefei, China, June 2010.
[87] A. Kwasinski and P.T. Krein, Optimal Configuration Analysis of a Microgrid-Based Telecom Power System, in *Rec. INTELEC 2006*, pp. 602–609, 2006.
[88] NREL, PV Solar Radiation, http://www.nrel.gov/gis/images/map_pv_us_september_may2004.jpg [Accessed December 27, 2009].
[89] Ettus Research. http://www.ettus.com/.
[90] Gnu Radio. http://gnuradio.org/.
[91] Mobile Ad-Hoc Networks Charter. http://datatracker.ietf.org/wg/manet/charter/.

[92] NSF Workshop on Distributed Processing over Cognitive Networks. http://www.ee.ucla.edu/NSF%20workshop%202009.htm.

[93] Et Docket No 03-222 Notice of Proposed Rulemaking and Order. Technical Report, FCC, December 2003.

[94] K. Akabane, H. Shiba, M. Matsui, and K. Uehara. An Autonomous Adaptive Base Station That Supports Multiple Wireless Network Systems. In *2nd IEEE International Symposium on New Frontiers in Dynamic Spectrum Access Networks (DySPAN)*, pp. 85–88, April 2007.

[95] Ian F. Akyildiz, Won-Yeol Lee, Mehmet C. Vuran, and Shantidev Mohanty. Next Generation/Dynamic Spectrum Access/Cognitive Radio Wireless Networks: A Survey. *Computer Networks*, vol. 50, no. 1, pp. 2127–2159, 2006.

[96] Ian F. Akyildiz, Xudong Wang, and Weilin Wang. Wireless Mesh Networks: A Survey. *Computer Networks*, vol. 47, no. 4, pp. 445–487, 2005.

[97] Albert-Lszl Barabsi, Albert laszlo Barabasi, Reka Albert, and Hawoong Jeong. Mean-field Theory for Scale-Free Random Networks, 1999.

[98] C. Bettstetter. Self-Organization in Computer and Communication Networks. http://www.bettstetter.com/.

[99] V. Brik, E. Rozner, S. Banerjee, and P. Bahl. DSAP: A Protocol for Coordinated Spectrum Access. In *First IEEE International Symposium on New Frontiers in Dynamic Spectrum Access Networks (DySPAN)*, pp. 611–614, November 2005.

[100] J.M. Celentano. Carrier Capital Expenditures. *Communications Magazine, IEEE*, vol. 46, no. 7, pp. 82–88, July 2008.

[101] International Engineering Consortium. Cellular Communications Tutorial, 2003.

[102] S. Dixit and A. Sarma. Advances in Self-Organizing Networks. *Communications Magazine, IEEE*, vol. 43, no. 7, pp. 76-77, July 2005.

[103] S. Dixit, E. Yanmaz, and O.K. Tonguz. On the Design of Self-Organized Cellular Wireless Networks. *Communications Magazine, IEEE*, vol. 43, no. 7, pp. 86-93, July 2005.

[104] R. Etkin, A. Parekh, and D. Tse. Spectrum Sharing for Unlicensed Bands. *IEEE Journal on Selected Areas in Communications*, vol. 25, no. 3, pp. 517–528, April 2007.

[105] Michel Bousquet and Gerard Maral. *Satellite Communications Systems – Systems, Techniques and Technology*. Wiley, 2009.

[106] A. Ghasemi and E.S. Sousa. Collaborative Spectrum Sensing for Opportunistic Access in Fading Environments. In *First IEEE International Symposium on New Frontiers in Dynamic Spectrum Access Networks (DySPAN)*, pp. 131–136, November 2005.

[107] S. Haykin. Cognitive Radio: Brain-Empowered Wireless Communications. *IEEE Journal on Selected Areas in Communications*, vol. 23, no. 2, pp. 201–220, February 2005.

[108] O. Holland, M. Muck, P. Martigne, D. Bourse, P. Cordier, S. Ben Jemaa, P. Houze, D. Grandblaise, C. Klock, T. Renk, Jianming Pan, P. Slanina, K. Mobner, L. Giupponi, J.P. Romero, R. Agusti, A. Attar, and A.H. Aghvami. Development of a Radio Enabler for Reconfiguration Management within the IEEE P1900.4 Working Group. In *2nd IEEE International Symposium on New Frontiers in Dynamic Spectrum Access Networks (DySPAN)*, pp. 232–239, April 2007.

[109] J. Huang, R.A. Berry, and M.L. Honig. Spectrum Sharing with Distributed Interference Compensation. In *First IEEE International Symposium on New Frontiers in Dynamic Spectrum Access Networks (DySPAN)*, pp. 88–93, November 2005.

[110] F.K. Jondral. Cognitive Radio: A Communications Engineering View. *Wireless Communications, IEEE*, vol. 14, no. 4, pp. 28–33, August 2007.

[111] V. Kanodia, A. Sabharwal, and E. Knightly. MOAR: A Multi-Channel Opportunistic Autorate Media Access Protocol for Ad Hoc Networks. In *First International Conference on Broadband Networks (BROADNETS)*, pp. 600-610, October 2004.

[112] Jeffrey O. Kephart and David M. Chess. The Vision of Autonomic Computing. *Computer*, vol. 36, pp. 41–50, 2003.

[113] R. Litjens, A. Eisenbltter, M. Amirijoo, O. Linnell, C. Blondia, T. Krner, N. Scully, L.C. Schmelz, J.L. van den Berg, and J. Oszmianski. Self-Configuration, Optimisation and Healing in Wireless Networks. *Wireless World Research Forum Meeting 20*, 2008.

[114] Won-Yeol Lee and I.F. Akyildiz. Optimal Spectrum Sensing Framework for Cognitive Radio Networks. *IEEE Transactions on Wireless Communications*, vol. 7, no. 10, pp. 3845–3857, October 2008.

[115] L. Ma, X. Han, and C.-C. Shen. Dynamic Open Spectrum Sharing Mac Protocol for Wireless Ad Hoc Networks. In *First IEEE International Symposium on New Frontiers in Dynamic Spectrum Access Networks (DySPAN)*, pp. 203–213, November 2005.

[116] N. Marchetti, N.R. Prasad, J. Johansson, and Tao Cai. Self-Organizing Networks: State-of-the-Art, Challenges and Perspectives. In *8th International Conference on Communications (COMM)*, pp. 503–508, June 2010.

[117] Rekha Menon, R.M. Buehrer, and J.H. Reed. Outage Probability Based Comparison of Underlay and Overlay Spectrum Sharing Techniques. In *First IEEE International Symposium on New Frontiers in Dynamic Spectrum Access Networks (DySPAN)*, pp. 101-109, November 2005.

[118] Shridhar Mubaraq Mishra, Anant Sahai, and Robert W. Brodersen. Cooperative Sensing among Cognitive Radios. In *Proc. of the IEEE International Conference on Communications (ICC)*, pp. 1658–1663, 2006.

[119] J. Mitola III and G.Q. Maguire Jr., Cognitive Radio: Making Software Radios More Personal. *Personal Communications, IEEE*, vol. 6, no. 4, pp. 13–18, August 1999.

[120] N. Nie and C. Comaniciu. Adaptive Channel Allocation Spectrum Etiquette for Cognitive Radio Networks. In *First IEEE International Symposium on New Frontiers in Dynamic Spectrum Access Networks (DySPAN)*, pp. 269-278, November 2005.

[121] Patricia A. Jacobs and P. Gaver. Planning Service to Provide Disaster Relief: Generic Command and Control Models. Technical Report, The Hastily Formed Networks (HFN) Research Group, 2007.

[122] P. Papadimitratos, S. Sankaranarayanan, and A. Mishra. A Bandwidth Sharing Approach to Improve Licensed Spectrum Utilization. *Communications Magazine, IEEE*, vol. 43, no. 12, Suppl. 10–14, December 2005.

[123] N. Pratas, N. Marchetti, N.R. Prasad, A. Rodrigues, and R. Prasad. Centralized Cooperative Spectrum Sensing for Ad-Hoc Disaster Relief Network Clusters. In *IEEE International Conference on Communications (ICC)*, pp. 1-5, May 2010.

[124] Nuno Pratas, Nicola Marchetti, Neeli Rashmi Prasad, Antonio Rodrigues, and Ramjee Prasad. Decentralized Cooperative Spectrum Sensing for Ad-Hoc Disaster Relief Net-

work Clusters. In *IEEE Vehicular Technology Conference (VTC-Spring)*, pp. 1–5, May 2010.

[125] C. Prehofer and C. Bettstetter. Self-Organization in Communication Networks: Principles and Design Paradigms. *Communications Magazine, IEEE*, vol. 43, no. 7, pp. 78–85, July 2005.

[126] Zhi Quan, Shuguang Cui, H. Poor, and A. Sayed. Collaborative Wideband Sensing for Cognitive Radios. *Signal Processing Magazine, IEEE*, vol. 25, no. 6, pp. 60–73, November 2008.

[127] J.M. Rabaey. A Brand New Wireless Day. In *Ambient Society Symposium (ASPDAC)*, p. 1, March 2008.

[128] C. Raman, R.D. Yates, and N.B. Mandayam. Scheduling Variable Rate Links via a Spectrum Server. In *First IEEE International Symposium on New Frontiers in Dynamic Spectrum Access Networks (DySPAN)*, pp. 110–118, November 2005.

[129] J. Rattner. Crossing the Chasm between Humans and Machines ... The Next 40 Years. Intel Developer Forum, 2008.

[130] H. Singh Mehta and S.A. Zekavat. Dynamic Resource Allocation via Clustered MC-CDMA in Multi-Service Ad-Hoc Networks: Achieving Low Interference Temperature. In *2nd IEEE International Symposium on New Frontiers in Dynamic Spectrum Access Networks (DySPAN)*, pp. 266–269, April 2007.

[131] J.A. Stine. A Location-Based Method for Specifying RF Spectrum Rights. In *2nd IEEE International Symposium on New Frontiers in Dynamic Spectrum Access Networks (DySPAN)*, pp. 34–45, April 2007.

[132] A.P. Subramanian and H. Gupta. Fast Spectrum Allocation in Coordinated Dynamic Spectrum Access Based Cellular Networks. In *2nd IEEE International Symposium on New Frontiers in Dynamic Spectrum Access Networks (DySPAN)*, pp. 320–330, April 2007.

[133] Jack Unger. *Deploying License-Free Wireless Wide-Area Networks*. Cisco Press, 2003.

[134] H. Urkowitz. Energy Detection of Unknown Deterministic Signals. *Proceedings of the IEEE*, vol. 55, no. 4, pp. 523–531, April 1967.

[135] Pramod K. Varshney. *Distributed Detection and Data Fusion*. Springer-Verlag, New York, 1996.

[136] E. Visotsky, S. Kuffner, and R. Peterson. On Collaborative Detection of TV Transmissions in Support of Dynamic Spectrum Sharing. In *First IEEE International Symposium on New Frontiers in Dynamic Spectrum Access Networks (DySPAN)*, pp. 338–345, November 2005.

[137] Xiao Fan Wang and Guanrong Chen. Complex Networks: Small-World, Scale-Free and Beyond. *Circuits and Systems Magazine, IEEE*, vol. 3, no. 1, pp. 6–20, 2003.

[138] S.A. Zekavat and X. Li. User-Central Wireless System: Ultimate Dynamic Channel Allocation. In *First IEEE International Symposium on New Frontiers in Dynamic Spectrum Access Networks (DySPAN)*, pp. 82–87, November 2005.

[139] Q. Zhang, F.W. Hoeksema, A.B.J. Kokkeler, and G.J.M. Smit. Towards Cognitive Radio for Emergency Networks. In *Mobile Multimedia: Communication Engineering Perspective*. Nova Publishers, USA, 2006.

[140] Jun Zhao, Haito Zheng, and Guang-Hua Yang. Distributed Coordination in Dynamic Spectrum Allocation Networks. In *First IEEE International Symposium on New Fron-*

tiers in Dynamic Spectrum Access Networks (DySPAN), pp. 259–268, November 2005.

[141] Q. Zhao, L. Tong, and A. Swami. Decentralized Cognitive MAC for Dynamic Spectrum Access. In *First IEEE International Symposium on New Frontiers in Dynamic Spectrum Access Networks (DySPAN)*, pp. 224–232, November 2005.

[142] H. Zheng and Lili Cao. Device-Centric Spectrum Management. In *First IEEE International Symposium on New Frontiers in Dynamic Spectrum Access Networks (DySPAN)*, pp. 56–65, November 2005.

[143] Haitao Zheng and Chunyi Peng. Collaboration and Fairness in Opportunistic Spectrum Access. *IEEE International Conference on Communications (ICC)*, vol. 5, pp. 3132-3136, May 2005.

[144] *Army Field Manual No. 7-8 – Infantry Rifle Platoon and Squad*, 1992, http://www.globalsecurity.org/military/library/policy/army/fm/7-8/.

[145] N. Aschenbruck, E. Gerhards-Padilla, and P. Martini, A Survey on Mobility Models for Performance Analysis in Tactical Mobile Networks. *Journal of Telecommunications and Information Technology (JTIT)*, vol. 2, pp. 54–61, 2008.

[146] N. Aschenbruck, E. Gerhards-Padilla, and P. Martini, Modeling Mobility in Disaster Area Scenarios. *Special Issue on Performance Evaluation of Wireless Ad Hoc, Sensor and Ubiquitous Networks Elsevier Performance Evaluation*, vol. 66, no. 12, pp. 773–790, 2009.

[147] N. Aschenbruck, M. Gerharz, M. Frank, and P. Martini, Modelling Voice Communication in Disaster Area Scenarios. In *Proc. IEEE Conf. on Local Computer Networks (LCN2006)*, pp. 211–220, 2006.

[148] N. Aschenbruck, M. Gerharz, and P. Martini, How to Assign Traffic Sources to Nodes in Disaster Area Scenarios. In *Proc. of the 26th IEEE Intern. Performance Computing and Communications Conf. (IPCCC)*, 2007.

[149] N. Aschenbruck and P. Martini, Evaluation and Parameterization of Voice Traffic Models for Disaster Area Scenarios. In *Proc. IEEE Conf. on Local Computer Networks (LCN2008)*, pp. 236–243, 2008.

[150] N. Aschenbruck, P. Martini, and M. Gerharz, Characterisation and Modelling of Voice Traffic in First Responder Networks. In *Proc. IEEE Conf. on Local Computer Networks (LCN2007)*, pp. 295–302, 2007.

[151] F. Bai and A. Helmy, A Survey of Mobility Models, http://nile.usc.edu/ helmy/important/Modified-Chapter1-5-30-04.pdf, 2004.

[152] C. Bettstetter, Mobility Modeling in Wireless Networks: Categorization, Smooth Movement, and Border Effects, *ACM SIGMOBILE Mobile Computing and Communications Review*, vol. 5, no. 3, pp. 55–66, 2001.

[153] P.T. Brady, A Model for Generating On-Off Speech Patterns in Two-Way Conversation, *The Bell System Technical Journal*, vol. 48, no. 9, pp. 2445–2472, 1969.

[154] J. Broch, D.A. Maltz, D.B. Johnson, Y.-C. Hu, and J. Jetcheva, A Performance Comparision of Multi-Hop Wireless Ad Hoc Network Routing Protocols. In *Proc. of the ACM/IEEE Mobicom*, pp. 85–97, 1998.

[155] T. Camp, J. Boleng, and V. Davies, A Survey of Mobility Models for Ad Hoc Network Research. *Wireless Communication and Mobile Computing (WCMC): Special issue on Mobile Ad Hoc Networking: Research, Trends and Applications*, vol. 2, no. 5, pp. 483–502, September 2002.

[156] G. Campos and G. Elias, Performance Issues of Ad Hoc Routing Protocols in a Network Scenario Used for Videophone Applications. In *Proceedings of the 38th Annual Hawaii International Conference on System Sciences*, 2005.

[157] *The Network Simulator – ns-2*, CMU Monarch, http://www.isi.edu/nsnam/ns/.

[158] J.S. Collura and D.J. Rahikka, Interoperable Secure Voice Communications in Tactical Systems. In *IEE Colloquium on Speech Coding Algorithms for Radio Channels*, 2000.

[159] S.R. Das, C.E. Perkins, and E.M. Royer, Performance Comparision of Two On-Demand Routing Protocols for Ad Hoc Networks. In *Proc. of the Infocom*, pp. 3–12, 2000.

[160] R. Ernst, N. Aschenbruck, and P. Martini, A Tactical Indoor Mobility Model for Urban Warfare Scenarios, submitted for publication, 2010.

[161] Z. Fu, H. Luo, P. Zerfos, S. Lu, L. Zhang, and M. Gerla, The Impact of Multihop Wireless Channel on TCP Performance, *IEEE Transactions On Mobile Computing*, vol. 4, no. 2, pp. 209–221, 2005.

[162] M. Gerharz, C. de Waal, M. Frank, and P. James, A Practical View on Quality-of-Service Support in Wireless Ad Hoc Networks. In *Proc. of the IEEE Workshop on Applications and Services in Wireless Networks (ASWN)*, pp. 185–196, 2003.

[163] J. Harri, F. Filali, and C. Bonnet, Mobility Models for Vehicular Ad Hoc Networks: A Survey and Taxonomy, Research Report RR-06-168, 2006.

[164] N. Haslett and A. Bonney, Loading Considerations for Public Safety Dispatch on Trunked Radio Systems. In *37th IEEE Vehicular Technology Conference (VTC)*, pp. 24–31, 1987.

[165] G. Hess and J. Cohn, Communications Load and Delay in Mobile Trunked Systems. In *Proceedings of IEEE Vehicular Technology Conference*, pp. 269–273, 1981.

[166] X. Hong, M. Gerla, G. Pei, and C.-C. Chiang, A Group Mobility Model for Ad Hoc Wireless Networks. In *Proceedings of the ACM International Workshop on Modelling and Simulation of Wireless and Mobile Systems (MSWiM)*, pp. 53–60, 1999.

[167] S.P. Hoogendoorn and P.H.L. Bovy, State-of-the-Art of Vehicular Traffic Flow Modelling, *Journal of Systems and Control Engineering – Special Issue on Road Traffic Modelling and Control*, vol. 215, no. 4, pp. 283–304, 2001.

[168] Y. Huang, W. He, K. Nahrstedt, and W.C. Lee, Corps: Event-Driven Mobility Model for First Responders in Incident Scene. In *Proceedings of the IEEE Military Communications Conference*, pp. 1–7, 2008.

[169] P.59: Artificial Conversational Speech, International Telecommunication Union – Telecommunication Standardization Sector, http://www.itu.int/rec/T-REC-P.59/, 1993.

[170] M. Jahnke, A. Wenzel, and G. Klein, FKIE-Bericht Nr. 163: Verfahren zur Erkennung von Angriffen gegen taktische MANETs, Forschungsinstitut für Kommunikation, Informationsverarbeitung und Ergonomie (FGAN-FKIE), Technical Report, 2008.

[171] P. Johansson, T. Larsson, N. Hedman, B. Mielczarek, and M. Degermark, Scenario-Based Performance Analysis of Routing Protocols for Mobile Ad-hoc Networks. In *Proceedings of the Mobicom*, pp. 195–206, 1999.

[172] D. B. Johnson and D. A. Maltz, Dynamic Source Routing in Ad Hoc Wireless Networks, *Mobile Computing*, vol. 353, pp. 153–181, 1996.

[173] J. Jordán and F. Barceló, Statistical Modelling of Channel Occupancy in Trunked PAMR Systems. In *Proceedings of 15th International Teletraffic Conference*, pp. 1169–1178, 1997.

[174] J. Jordán and F. Barceló, Statistical Modelling of Transmission Holding Time in PAMR Systems. In *Proceedings IEEE Globecom*, pp. 121–125, 1997.

[175] S. Karpinski, E.M. Belding-Royer, and K.C. Almeroth, Wireless Traffic: The Failure of CBR Modeling. In *Proceedings of IEEE International Conference on Broadband Communications, Networks and Systems*, 2007.

[176] W.-H. Liao, Y.-C. Tseng, and J.-P. Sheu, GRID: A Fully Location-Aware Routing Protocol for Mobile Ad Hoc Networks. *Telecommunication Systems*, vol. 18, no. 1–3, pp. 37–60, 2001.

[177] P. MESA, Making Progress toward an International PPDR Standard, www.projectmesa.org/whitepaper/MESA_whitepaper.pdf, 2003.

[178] S. Reidt and S.D. Wolthusen, An Evaluation of Cluster Head TA Distribution Mechanisms in Tactical MANET Environments. In *Proceedings of the Annual Conference of ITA (ACITA)*, 2007.

[179] S. Reidt and S.D. Wolthusen, Efficient Trust Authority Distribution in Tactical MANET Environments. *Proceedings of the IEEE Military Communications Conference (MILCOM)*, 2007.

[180] P. Santi, Topology Control in Wireless Ad Hoc and Sensor Networks, *ACM Computer Surveys*, vol. 37, no. 2, pp. 164–194, June 2005.

[181] D.S. Sharp, N. Cackov, N. Lasković, Q. Shao, and L. Trajković, Analysis of Public Safety Traffic on Trunked Land Mobile Radio Systems. *IEEE J-SAC*, vol. 22, no. 7, pp. 1197–1205, 2004.

[182] G.M. Stone, Public Safety Wireless Communications User Traffic Profiles and Grade of Service Recommendations, U.S. Department of Justice, SRSC Final Report, Appendix D, 1996.

[183] K. Sundaresan, H. Hsieh, and R. Sivakumar, IEEE 802.11 Over Multi-Hop Wireless Networks: Problems and New Perspectives. *Ad Hoc Networks*, vol. 2, no. 2, pp. 109–132, 2004.

[184] J. Tian, J. Hähner, C. Becker, I. Stepanov, and K. Rothermel, Graph-based Mobility Model for Mobile Ad Hoc Network Simulation. In *Proceedings of the 35th Annual Simulation Symposium*, pp. 337–344, 2002.

[185] Y.-C. Tseng, S.-Y. Ni, Y.-S. Chen, and J.-P. Sheu, The Broadcast Storm Problem in a Mobile Ad Hoc Network, *Wireless Networks*, vol. 8, no. 2-3, pp. 153–167, 2002.

[186] B. Vujicić, N. Cackov, S. Vujicić, and L. Trajković, Modeling and Characterization of Traffic in Public Safety Wireless Networks, *Proc. SPECTS*, pp. 214–223, 2005.

[187] S. Xu and T. Saadawi, Does the IEEE 802.11 MAC Protocol Work Well in Multihop Wireless Ad Hoc Networks? *IEEE Communications Magazine*, vol. 39, no. 6, pp. 130–137, 2001.

[188] P. Yao, E. Krohne, and T. Camp, Performance Comparision of Geocast Routing Protocols for a MANET. In *Proceedings of the 13th IEEE International Conference on Computer Communications and Networks (IC3N)*, pp. 213–220, 2004.

[189] Y. Yi, S.-J. Lee, W. Su, and M. Gerla, Internet-Draft: On-Demand Multicast Routing Protocol (ODMRP) for Ad Hoc Networks, IETF – Mobile Ad Hoc Networks Working Group, draft-ietf-manet-odmrp-04.txt, 2002.

[190] Emergency Communications during Natural Disasters: Infrastructure and Technology, http://tsunami.ait.ac.th/Documents/disaster_communication_assistance_concept_paper.pdf

[191] A. Boukalov, Cross Standard System for Future Public Safety and Emergency Communications. In *Proceedings of IEEE 60th Vehicular Technology Conference – Fall, 2004 (VTC2004-Fall)*, Los Angeles, USA, September, vol. 7, pp. 5224–5229, 2004.

[192] K. Okada, Limiting the Holding Time in Mobile Cellular Systems during Heavy Call Demand Periods in the Aftermath of Disasters, *IEICE Transactions Fundamental*, vol. E85-A, no. 7, pp. 1454–1462, July 2002.

[193] G. Sato et al., A Combination of Different Wireless LANs to Realize Disaster Communication Network. In *Proceedings of IEEE International Conference on Distributed Computing Systems Workshops (ICDCS Workshops '09)*, 22–26 June 2009, pp. 395–399, 2009.

[194] F.R. Yu et al., Enhancing Interoperability in Heterogeneous Mobile Wireless Networks for Disaster Response. *IEEE Transactions on Wireless Communications*, vol. 8, no. 5, pp. 2424–2433, May 2009

[195] G. Iapichino et al., A Mobile Ad Hoc Satellite and Wireless Mesh Networking Approach for Public Safety Communications. In *Proceedings of 10th IEEE International Workshop on Signal Processing for Space Communications (SPSC 2008)*, October 2008, pp. 1–6, 2008.

[196] K. Kanchanasut et al., DUMBONET: A Multimedia Communications System for Collaborative Emergency Response Operation in Disaster-Affected Areas. *International Journal of Emergency Management (IJEM)*, vol. 4, no. 4, pp. 670–681, 2007.

[197] H. Suzuki et al., An Ad Hoc Network in the Sky, SKYMESH, for Large-Scale Disaster Recovery. In *Proceedings of IEEE Vehicular Technology Conference (VTC-2006 Fall)*, September 2006, pp. 1–5, 2006.

[198] M. Kimura et al., A Study on Emergency Multi-System Access in Mobile Cellular Systems during Large Scale Disasters, IEICE Technical Report, CQ2004-51 (2004-07), pp. 12–18, July 2008.

[199] H.N. Nguyen et al., Emergency Multi-System Access with Selecting Proper Base Station Utilizing WCDMA Networks for Emergency Communications. In *Proceedings of International Conference on Wireless Communication, Vehicular Technology, Information Theory and Aerospace & Electronic Systems Technology (Wireless VITAE 2009)*, May 2009, pp. 530–534, 2009.

[200] http://www.d2c.co.jp/library/profile/news/2009/mobile_0902e.pdf

[201] H. Holma and A. Toskala, *WCDMA for UMTS: Radio Access for Third Generation Mobile Communications*, John Wiley & Sons Publisher, 2004.

[202] ARIB Evaluation Group, Evaluation Methodology IMT-2000 Radio Transmission Technologies (Version 1.1), September 1999.

[203] N. Jesuale and Bernard C. Eydt, A Policy Proposal to Enable Cognitive Radio for Public Safety and Industry in the Land Mobile Radio Bands. In *New Frontiers in Dynamic Spectrum Access Networks, DySPAN 2007, 2nd IEEE International Symposium*, Dublin, Ireland, 17–20 April, pp. 66–77, 2007.

[204] National Public Safety Telecommunications Council. [Online]. Available: http://www.npstc.org/npstcintro.jsp.

[205] Federal Communications Commission, Public Safety and Homeland Security Bureau, 700 MHz Public Safety Spectrum. [Online]. Available: http://www.fcc.gov/pshs/spectrum/700-MHz/.

[206] J. Mitola III and G.Q. Maguire, Jr., Cognitive Radio: Making Software Radios More Personal. *IEEE Personal Communications Mag.*, vol. 6, no. 4, pp. 13–18, August 1999.

[207] H. Arslan and J. Mitola III, Guest Editorial. Special Issue: Cognitive Radio, Software Defined Radio, and Adaptive Wireless Systems, *Wireless Communication and Mobile Computing Journal*, vol. 7, no. 9, pp. 1333–1335, 2007.

[208] D. Cabric, S.M. Mishra, and R.W. Brodersen, Implementation Issues in Spectrum Sensing for Cognitive Radios. In *Proceedings IEEE Asilomar Conference on Signals, Systems, Computers*, Pacific Grove, CA, USA, vol. 1, pp. 772–776, November 2004.

[209] G.P. Fettweis, K. Iversen, M. Bronzel, H. Schubert, V. Aue, D. Maempel, J. Voigt, A. Wolisz, G. Walf, and J.P. Ebert, A Closed Solution for an Integrated Broadband Mobile System (IBMS). In *Proceedings of International Conference on Universal Personal Communications (ICUPC'96)*, Cambridge, Massachusetts, USA, pp. 707–711, October 1996.

[210] M. Bronzel, D. Hunold, G. Fettweis, T. Konschak, T. Doelle, V. Brankovic, H. Alikhani, J.-P. Ebert, A. Festag, F. Fitzek, and A. Wolisz, Integrated Broadband Mobile System (IBMS) featuring Wireless ATM. In *Proceedings of ACTS Mobile Communication Summit 97*, Aalborg, Denmark, pp. 641–646, October 1997.

[211] Joseph Mitola III, Software Radios: Survey, Critical Evaluation and Future Directions. *IEEE Aerospace and Electronic Systems Magazine*, vol. 8, no. 4, pp. 25–36, April 1993.

[212] Joseph Mitola III, Cognitive Radio: An Integrated Agent Architecture for Software Defined Radio, Ph.D. Dissertation, KTH Royal Institute of Technology, Stockholm, Sweden, May 8, 2000.

[213] Federal Communications Comission, Technical Report. In *Et Docket No. 03-322. Notice of Proposed Rule Making and Order*, Washington, DC , December 2003.

[214] W. Krenik and A. Batra, Cognitive Radio Techniques for Wide Area Networks. In *Proceedings ACM IEEE Design Automation Conference 2005*, Anaheim, CA, USA, pp. 409–412, June 2005.

[215] R.W. Brodersen, A. Wolisz, D. Cabric, S.M. Mishra, and D. Willkomm, Corvus: A Cognitive Radio Approach for Usage of Virtual Unlicensed Spectrum. White Paper, Berkeley Wireless Research Center (BWRC), CA, 2004.

[216] M.J. Marcus, Unlicensed Cognitive Sharing of TV Spectrum: The Controversy at the Federal Communications Commission. *IEEE Commun. Mag.*, vol. 43, pp. 24–25, May 2005.

[217] A. Sahai, N. Hoven, and R. Tandra, Some Fundamental Limits on Cognitive Radio. In *Proceedings of Allerton Conference on Communication, Contol and Computing*, Monticello, October 2004.

[218] Haiyun Tang, Some Physical Layer Issues of Wide-band Cognitive Radio Systems. In *New Frontiers in Dynamic Spectrum Access Networks (DySPAN 2005)*, Baltimore, Maryland, USA, pp. 151–159, November 2005.

[219] R. Murty, Software-Defined Reconfigurable Radios: Smart, Agile, Cognitive, and Interoperable. *Technology at Intel Magazine,* June 2003.

[220] D. Scaperoth, B. Le, T.W. Rondeau, D. Maldonado, and C.W. Bostian, Cognitive Radio Platform Development for Interoperability. In *Proceedings Military Communications Conference (MILCOM'06)*, Washington DC, USA, October 2006.

[221] John S. Powell, Cognitive and Software Radio: A Public Safety Regulatory Perspective. In *National Public Safety Telecommunications Council Committee and Governing Board Meeting*, Washington DC, 14–15 June 2004.

[222] Federal Communications Commission, Report to Congress, On the Study to Assess Short-Term and Long-Term Needs for Allocations of Additional Portions of the Electromagnetic Spectrum for Federal, State and Local Emergency Response Providers. [Online]. Available: http://fjallfoss.fcc.gov/edocs public/attachmatch/DOC-262865A1.pdf.

[223] U.S. Department of Homeland Security, Public Safety Statement of Requirements for Communications and Interoperability, vol. 1, ver. 1.2, October 2006. [Online]. Available: http://www.safecomprogram.gov.

[224] SDR Forum, SDR Technology for Public Safety Report, 2006. [Online]. Available: www.ece.vt.edu/swe/chamrad/psi/SDRF-06-A-0001-V0.00.pdf.

[225] S.D. Jones, E. Jung, Xin Liu, N. Merheb, and I-Jeng Wang, Characterization of Spectrum Activities in the U.S. Public Safety Band for Opportunistic Spectrum Access. In *New Frontiers in Dynamic Spectrum Access Networks (DySPAN 2007), 2nd IEEE International Symposium*, Dublin, Ireland, 17–20 April, pp. 137–146, 2007.

[226] Terrestrial Trunked Radio (TETRA) Association. [Online]. Available: http://www.tetra-association.com/.

[227] Emergency Ultrawideband Radio for Positioning and Communications. [Online]. Available: http://www.ist-europcom.org/.

[228] Mobile Broadband for Public Safety Project. [Online]. Available: http://www.projectmesa.org/.

[229] Cognitive Radio Systems for First Responders. [Online]. Available: http://www.sharedspectrum.com/press/pdf/pr101507.pdf.

[230] S.M. Hasan, P. Balister, K. Lee, J. Reed, and S. Ellingson, A Low Cost Multi-band/Multi-mode Radio for Public Safety Applications. In *2006 Software Defined Radio Forum (SDR Forum) Technical Conference*, Vancouver, BC, November 2006.

[231] Bin Le, Francisco A.G. Rodriguez, Qinqin Chen, Bin Philip Li, Feng Ge, Mustafa El-Nainay, Thomas W. Rondeau, and Charles W. Bostian, A Public Safety Cognitive Radio Node. In *2007 Software Defined Radio Forum (SDR Forum) Technical Conference*, Denver, CO, USA, November 2007.

[232] N. Jesuale, Overview of State and Local Government Interests in Spectrum Policy Issues. In *New Frontiers in Dynamic Spectrum Access Networks (DySPAN 2005)*, Baltimore, Maryland, USA, pp. 476–485, November 2005.

[233] F. Hoeksema, M. Heskamp, R. Schiphorst, and K. Slump, A Node Architecture for Disaster Relief Networking. In *New Frontiers in Dynamic Spectrum Access Networks (DySPAN 2005)*, Baltimore, Maryland, USA, November, pp. 577–584, 2005.

[234] P. Pawelczak, R. Venkatesha Prasad, L. Xia, and I.G.M.M. Niemegeers, Cognitive Radio EmNrgency networks – Requirements and Design. In *New Frontiers in Dynamic Spectrum Access Networks (DySPAN 2005)*, Baltimore, Maryland, USA, pp. 601–606, November 2005.

[235] Qiwei Zhang, A.B.J. Kokkeler, and G.J.M. Smit, A Reconfigurable Radio Architecture for Cognitive Radio in Emergency Networks. In *Proceedings of the 9th European Conference on Wireless Technology*, Manchester, UK, pp. 35–38, September 2006.

[236] P.J. Green and D.P. Taylor, A Real Time Cognitive Radio Test Platform for Public Safety Physical Layer Experiments. In *Proceedings of IEEE 18th International Symposium on Personal, Indoor and Mobile Radio Communications (PIMRC 2007)*, Athens, Greece, 3–7 September, pp. 1–5, 2007.

[237] Wei Wang, Weidong Gao, Xinyu Bai, Tao Peng, Gang Chuai, and Wenbo Wang, A Framework of Wireless Emergency Communications Based on Relaying and Cognitive Radio. In *Proceedings of IEEE 18th International Symposium on Personal, Indoor and Mobile Radio Communications (PIMRC 2007)*, Athens, Greece, 3–7 September, pp. 1–5, 2007.

[238] S. Majid and K. Ahmed, Self-Recognition Emergency Communications for Mobile Handsets. In *Proceedings of Asia-Pacific Conference on Communications (APCC 2007)*, Bangkok, Thailand, 18–20 October, pp. 419–422, 2007.

[239] S. Majid and K. Ahmed, CIP – Cognitive Identification of Post-Disaster Communications. In *Proceedings of International Symposium on Communications and Information Technologies (ISCIT '07)*, Sydney, Australia, 16–19 October, pp. 763–767, 2007.

[240] Federal Communications Commission, 700 MHz Partnership. [Online]. Available: http://www.fcc.gov/pshs/spectrum/700mhz/partnership.html.

[241] M. Hata, Emprical Formula for Propagation Loss in Land Mobile Radio Services. *IEEE Trans. Veh. Tech.*, vol. 29, May 1980.

[242] Gerard K. Rauwerda, Jordy Potman, Fokke W. Hoeksema, and Gerard J.M. Smith, Adaptation in the Physical Layer Using Heterogeneous Reconfigurable Hardware. In *Adaptation Techniques in Wireless Multimedia Networks*, 2006.

[243] H. Celebi and H. Arslan, Cognitive Positioning Systems. *IEEE Trans. on Wireless Communications*, vol. 6, no. 12, pp. 4475–4483, December 2007.

[244] H. Celebi and H. Arslan, Ranging Accuracy in Dynamic Spectrum Access Networks. *IEEE Communications Letters*, vol. 11, no. 5, pp. 405–407, May 2007.

[245] H. Celebi and H. Arslan, Utilization of Location Information in Cognitive Wireless Networks. *IEEE Wireless Communications Magazine – Special Issue on Cognitive Wireless Networks*, vol. 14, no. 4, pp. 6–13, August 2007.

[246] H. Celebi and H. Arslan, Enabling Location and Environment Awareness in Cognitive Radios, Invited Paper. *Elsevier Computer Communications – Special Issue on Advanced Location-Based Services*, 2008.

[247] Serhan Yarkan and Huseyin Arslan, Exploiting Location Awareness towards Improved Wireless System Design in Cognitive Radio. *IEEE Communications Magazine, Feature Topic on Cognitive Radios for Dynamic Spectrum Access*, accepted for publication.

[248] H. Celebi and H. Arslan, Adaptive Positioning Systems for Cognitive Radios. In *Proceedings IEEE Symposium on New Frontiers in Dynamic Spectrum Access Networks (DySpan'07)*, Dublin, Ireland, April 17–20, pp. 78–84, 2007.

[249] Signal Intelligence (SIGINT) definition. [Online]. Available: http://en.wikipedia.org/wiki/SIGINT.

[250] Robert I. Desourdis Jr., David R. Smith, Richard J. Dewey, and John R. DiSalvo, *Emerging Public Safety Wireless Communication Systems*, 1st edition. Artech House Publishers, 2001.

[251] Summitek Co. Signal Intelligence Concept. [Online]. Available: www.summitekinstruments.com/oasis/docs/PR/SpectrumMon-SIGINT030106.pdf.

[252] Carlos Y. Aguayo Gonzalez, Francisco Porthelinha, and Jeff Reed, Design and Implementation of an SCA Core Framework for a DSP. In *2006 Software Defined Radio Forum (SDR Forum) Technical Conference*, Vancouver, BC, November 2006.

[253] Software Communications Architecture (SCA). [Online]. Available: http://sca.jpeojtrs.mil/downloads.asp?ID=2.2.2.

[254] G. Lind and C. Littke, Software Communication Architecture (SCA) for Above 2 GHz SATCOM. In *Proceeding of the SDR 04 Technical Conference and Product Exposition*, Phoenix, Arizona, USA, November 2004.

[255] Federal Communications Commission, Wireless 911 Services. [Online]. Available: http://www.fcc.gov/cgb/consumerfacts/wireless911srvc.html.

[256] C. Drane, M. Macnaughtan, and C. Scott, Positioning GSM Telephones. *IEEE Commun. Mag.*, vol. 36, no. 4, pp. 46–54, 1998.

[257] A. Gorcin, RSS-based Location Awareness for Public Safety Cognitive Radio. In *Proceedings IEEE Wireless Communications, Vehicular Technology, Information Theory and Aerospace & Electronic Systems Technology Conference (Wireless VITAE'09)*, Aalborg, Denmark, 17–20 May 2009.

[258] K. Pahlavan, X. Li, and J.P. Makela, Indoor Geolocation Science and Technology. *IEEE Commun. Mag.*, vol. 40, no. 2, pp. 112–118, 2002.

[259] R.A. Altes, A Theory for Animal Echolocation and Its Applications to Ultrasonics. In *Proceedings IEEE Ultrasonics Symposium*, Monterey, California, USA, November, pp. 67–72, 1993.

[260] T.K. Horiuchi, Seeing in the Dark: Neuromorphic VLSI Modeling of Bat Echolocation. *IEEE Signal Processing Magazine*, vol. 22, no. 5, pp. 134–139, 2005.

[261] D. Bank and T. Kampke, High-Resolution Ultrasonic Environment Imaging. *IEEE Trans. Robotics*, vol. 23, no. 2, pp. 370–381, 2007.

[262] H. Arslan, *Cognitive Radio, Software Defined Radio, and Adaptive Wireless Systems*. Springer, Berlin, 2007.

[263] S. Haykin, Cognitive Radio: Brain-Empowered Wireless Communications. *IEEE J. Select. Areas Commun.*, vol. 23, no. 2, pp. 201–220, 2005.

[264] M. Brunato and R. Battiti, Statistical Learning Theory for Location Fingerprinting in Wireless LANs. *Computer Networks*, vol. 47, no. 6, pp. 825–845, 2005.

[265] S. Capkun and J. Hubaux, Secure Positioning in Wireless Networks. *IEEE JSAC*, vol. 24, no. 2, pp. 221–32, 2006.

[266] H.-H. Nagel, Steps towards a Cognitive Vision System. *Artif. Intell. Mag.*, vol. 25, no. 2, pp. 31–50, 2004.

[267] R. Koch, Dynamic 3-D Scene Analysis through Synthesis Feedback Control. *IEEE Trans. Pattern Anal. Mach. Intell.*, vol. 15, no. 6, pp. 556–568, 1993.

[268] T. Starner, B. Schiele, and A. Pentland, Visual Context Awareness via Wearable Computing. In *Proceedings IEEE International Symposium on Wearable Computers*, Pittsburgh, PA, USA, pp. 50–57, October 1998.

[269] A. Quazi, B. Schiele, and A. Pentland, An Overview on the Time Delay Estimate in Active and Passive Systems for Target Localization. *IEEE Trans. Acoust. Speech, Signal Processing*, vol. 29, no. 3, pp. 527–533, 1981.

[270] E. Trevisani and A. Vitaletti, Cell-ID Location Technique, Limits and Benefits: An Experimental Study. In *Proceedings 6th IEEE Workshop on Mobile Computing Systems and Applications*, English Lake District, UK, pp. 51–60, December 2004.

[271] V. Zeimpekis, G. Giaglis, and G. Lekakos, A Taxonomy of Indoor and Outdoor Positioning Techniques for Mobile Location Services. *ACM SIGecom Exchanges*, vol. 3, no. 4, pp. 19–27, 2002.

[272] N. Patwari, A. Hero, M. Perkins, N. Correal, and R. O'Dea, Relative Location Estimation in Wireless Sensor Networks. *IEEE Trans. Signal Processing*, vol. 51, no. 8, pp. 2137–2148, August 2003.

[273] J. Hightower and G. Borriello, A Survey and Taxonomy of Location Systems for Ubiquitous Computing. Technical Report, University of Washington, 2001.

[274] S. Gezici, Z. Tian, G.B. Giannakis, H. Kobayashi, A.F. Molisch, H.V. Poor, and Z. Sahinoglu, Localization via UWB Radios. *IEEE Signal Processing Mag.*, vol. 22, no. 4, pp. 70–84, July 2005.

[275] Y. Qi and H. Kobayashi, On Relation among Time Delay and Signal Strength Based Geolocation Methods. *Proc. IEEE Globecom*, vol. 7, pp. 4079–4083, December 2003.

[276] StarFire: A Global High Accuracy Differential GPS System. [Online]. Available: http://www.navcomtech.com, 2006.

[277] A Taxonomy of Indoor and Outdoor Positioning Techniques for Mobile Location Services. [Online]. Available: www.acm.org, 2003,

[278] F. van Diggelen, Indoor GPS Theory & Implementation. In *Proceedings IEEE Position, Location and Navigation Symposium*, pp. 240–247, April 2002.

[279] Geolocation Development Document. [Online]. Available: http://www.ieee802.org/22/, 2006.

[280] Y. Xing, R. Chandramouli, S. Mangold, and S. Shankar, Dynamic Spectrum Access in Open Spectrum Wireless Networks. *IEEE J. Sel. Areas Commun.*, vol. 24, no. 3, pp. 626–637, March 2006.

[281] Y.-H. Lu and E.J. Delp, Image-Based Location Awareness and Navigation: Who Cares? In *Proceedings EEE Southwest Symposium on Image Analysis and Interpretation*, Lake Tahoe, Nevada, USA, pp. 26–30, March 2004.

[282] P. Luley, A. Almer, C. Seifert, G. Fritz, and L. Paletta, A Multi-Sensor System for Mobile Services with Vision Enhanced Object and Location Awareness. In *Proceedings IEEE International Mobile Commerce and Services*, Munich, Germany, pp. 52–59, July 2005.

[283] G. Sun, J. Chen, W. Guo, and K. Liu, Signal Processing Techniques in Network-Aided Positioning: A Survey of State-of-the-Art Positioning Designs. *IEEE Signal Process. Mag.*, vol. 22, no. 4, pp. 12–23, 2005.

[284] R. Murphy, Trial by Fire. *IEEE Robot. and Automat. Mag.*, vol. 11, no. 3, pp. 50–61, 2004.

[285] R. Murphy et al., Mobility and Sensing Demands in USAR. *IECON2000*, vol. 1, pp. 138–142, October 2000.

[286] K. Osuka, R. Murphy, and A. C. Schultz, USAR Competitions for Physically Situated Robots. *IEEE Robot. and Automat. Mag.*, vol. 9, no. 3, pp. 26–33, 2002.

[287] *IEEE International Workshop on Safety, Security, and Rescue Robotics (SSRR2008)*, Sendai, Japan, 2008.

[288] Session: Rough Terrain Mobility. In *Proceedings of the SSRR2008*, 2008.

[289] A. Kokosy, F. O. Defaux, and W. Perruquetti, Autonomous Navigation of a Nonholonomic Mobile Robot in a Complex Environment. In *Proceedings of the SSRR2008*, 2008.

[290] S.I. Ali and B. Mertsching, Towards a Generic Control Architecture of Rescue Robot Systems. In *Proceedings of the SSRR2008*, 2008.

[291] M. Trierscheid et al., Hyperspectral Imaging for Victim Detection with Rescue Robots. In *Proceedings of the SSRR2008*, 2008.

[292] A. Ollero et al., COMETS: A Multiple Heterogeneous UAV System. In *Proceedings of the SSRR '04*, 2004.

[293] L. Alboul, J. Saez-Pons, and J. Penders, Mixed Human-Robot Team Navigation in the GUARDIANS Project. In *Proceedings of the SSRR '08*, 2008.

[294] J. Vazquez and C. Malcolm, Distributed Multirobot Exploration Maintaining a Mobile Network. In *Proceedings of the IEEE IS'2004*, 2004.

[295] G. Antonelli et al., Use of a Robot Platoon to Implement Mobile Ad-hoc NeTwork in Rescue Scenario – Experimental Results. In *Proceedings of the SSRR2007*, 2007.

[296] C.E. Perkins and P. Bhagwat, Highly Dynamic Destination-Sequenced Distance-Vector Routing (DSDV) for Mobile Computers. In *ACM SIGCOMM'94*, pp. 234–244, 1994.

[297] C.C. Chiang, Routing in Clustered Multihop, Mobile Wireless Networks withFading Channel. In *Proceedings IEEE SICON' 97*, pp. 197–211, April 1997.

[298] C.E. Perkins and E.M. Royer, Ad-hoc On-Demand Distance Vector Routing. In *Proceedings 2nd IEEE Workshop Mobile Comp. Sys. and Apps.*, pp. 90–100, February 1999.

[299] D.B. Johnson and D.A. Maltz, Dynamic Source Routing in Ad-HocWireless Networks. In *Mobile Computing*, T. Imielinski and H. Korth (Eds.), Kluwer, pp. 153–181, 1996.

[300] H. Sugiyama, T. Tsujioka, and M. Murata, Ad Hoc Network Simulator Based on DSDV Routing Method. In *Memoirs of the Faculty of Engineering, Osaka City University*, pp. 43–58, 2002.

[301] H. Sugiyama, T. Tsujioka, and M. Murata, Autonomous Chain Network Formation by Multi-Robot Rescue System with Ad Hoc Networking. In *Proceedings of the SSRR '10*, 2010.

[302] H.G. Nguyen, H.R. Everett, N. Manouk, and A. Verma, Autonomous Mobile Communication Relays. In *SPIE Proceedings 4715: Unmanned Ground Vehicle Technology IV*, pp. 50–57, April 2002.

[303] W. Burgard et al., Coordinated Multi-Robot Exploration. *IEEE Trans. Robot. Autom.*, vol. 21, no. 2, pp. 376–386, June 2005.

[304] D. Fox, J. Ko, K. Konolige, B. Limketkai, D. Schulz, and B. Stewart, Distributed Multirobot Exploration and Mapping. *Proc. of IEEE*, vol. 94, no. 7, pp. 1325–1339, July 2006.

[305] A.S. Gavin and R.A. Brooks, *Low Computation Vision-Based Navigation for a Martian Rover*, American Institute of Aeronautics and Astronautics, 1993.

[306] F. Driewer et al., Hybrid Telematic Teams for Search and Rescue Operations. In *Proceedings of the SSRR '04*, 2004.

[307] S. Sukkarieh, E.M. Nebot, and H.F. Durrant-Whyte, A High Integrity IMU/GPS Navigation Loop for Autonomous Land Vehicle Applications. *IEEE Trans. on Robot. and Automat.*, vol. 15, no. 3, pp. 572–578, 1999.

[308] H. Sugiyama and M. Murata, AMRS: An Autonomous Mobile Robot Simulator. In *Memoirs of the Faculty of Engineering, Osaka City Univ.*, pp. 105–121, 1996.

[309] H. Sugiyama, T. Tsujioka, and M. Murata, QoS Routing in a Multi-Robot Network System for Urban Search and Rescue. In *Proceedings of the HWISE 2006*, pp. 323–327, 2006.

[310] H. Sugiyama, T. Tsujioka, and M. Murata, Improvement of Transmission Properties of Synchronized QoS Ad Hoc Network by Rapid Control Method. *Electronics and Communications in Japan: Part1*, vol. 89, no. 9, pp. 51–59, 2006.

[311] C.R. Lin, J.S. Liu, QoS Routing in Ad Hoc Wireless Network. *IEEE J. Select. Areas Commun.*, vol. 17, no. 8, pp. 1426–1438, 1999.

[312] G. Chen et al., Multi-hop Time Synchronization Protocol for IEEE 802.11 Wireless Ad Hoc Networks. *Journal of International Science and Engineering*, vol. 23, pp. 969–983, 2007.

[313] H. Sugiyama, T. Tsujioka, and M. Murata, Victim Detection System for Urban Search and Rescue Based on Active Network Operation. In *Design and Application of Hybrid Intelligent Systems*, IOS Press, Amsterdam, pp. 1104–1113, 2003.

[314] Coding of Audio-Visual Objects Part 2: Visual, AMD 1: Visual Extensions, *ISO/IEC* 14496-2:1999/Amd. 1:2000(E), January 2001.

Index

About the Editor

Dr. Nicola Marchetti is currently an Assistant Professor at the Networking and Security Group, Center for TeleInFrastruktur (CTIF) at Aalborg University, Denmark, where he has been working since January 2009. He received the Ph.D. in Wireless Communications and the M.Sc. in Mathematics from Aalborg University in the years 2007 and 2010 respectively, and the M.Sc. in Electronic Engineering from University of Ferrara, Italy in 2003. He worked as a Research Assistant at the University of Ferrara during July 2003 to April 2004. He also worked as a Ph.D. fellow at Aalborg University during May 2004 to May 2007.

Dr. Nicola Marchetti is the Co-Founder of the International Workshop on Cognitive Radio and Advanced Spectrum Management (CogART) and the International Symposium on Applied Sciences in Biomedical and Communication Technologies (ISABEL).

His research interests include: Multiple Antenna Technologies, Single & Multi Carrier Modulations, Advanced Radio Resource Management Techniques, Cognitive Radios and Networks, Self-organizing Networks, e-Healthcare, and Mathematics applied to Wireless Communication.

RIVER PUBLISHERS SERIES IN COMMUNICATIONS

Volume 1
4G Mobile & Wireless Communications Technologies
Sofoklis Kyriazakos, Ioannis Soldatos, George Karetsos
September 2008
ISBN: 978-87-92329-02-8

Volume 2
Advances in Broadband Communication and Networks
Johnson I. Agbinya, Oya Sevimli, Sara All, Selvakennedy Selvadurai, Adel Al-Jumaily, Yonghui Li, Sam Reisenfeld
October 2008
ISBN: 978-87-92329-00-4

Volume 3
Aerospace Technologies and Applications for Dual Use A New World of Defense and Commercial in 21st Century Security
General Pietro Finocchio, Ramjee Prasad, Marina Ruggieri
November 2008
ISBN: 978-87-92329-04-2

Volume 4
Ultra Wideband Demystified Technologies, Applications, and System Design Considerations
Sunil Jogi, Manoj Choudhary
January 2009
ISBN: 978-87-92329-14-1

Volume 5
Single- and Multi-Carrier MIMO Transmission for Broadband Wireless Systems
Ramjee Prasad, Muhammad Imadur Rahman, Suvra Sekhar Das, Nicola Marchetti
April 2009
ISBN: 978-87-92329-06-6

Volume 6
Principles of Communications: A First Course in Communications
Kwang-Cheng Chen
June 2009
ISBN: 978-87-92329-10-3

Volume 7
Link Adaptation for Relay-Based Cellular Networks
Başak Can
November 2009
ISBN: 978-87-92329-30-1

Volume 8
Planning and Optimisation of 3G and 4G Wireless Networks
J.I. Agbinya
January 2010
ISBN: 978-87-92329-24-0

Volume 9
Towards Green ICT
Ramjee Prasad, Shingo Ohmori, Dina Šimunić
June 2010
ISBN: 978-87-92329-38-7

Volume 10
Adaptive PHY-MAC Design for Broadband Wireless Systems
Ramjee Prasad, Suvra Sekhar Das, Muhammad Imadur Rahman
August 2010
ISBN: 978-87-92329-08-0

Volume 11
Multihop Mobile Wireless Networks
Kannan Govindan, Deepthi Chander, Bhushan G. Jagyasi, Shabbir N. Merchant, Uday B. Desai
October 2010
ISBN: 978-87-92329-44-8

For Product Safety Concerns and Information please contact our EU
representative GPSR@taylorandfrancis.com
Taylor & Francis Verlag GmbH, Kaufingerstraße 24, 80331 München, Germany

www.ingramcontent.com/pod-product-compliance
Ingram Content Group UK Ltd.
Pitfield, Milton Keynes, MK11 3LW, UK
UKHW021118180425
457613UK00005B/137

* 9 7 8 8 7 9 2 3 2 9 4 8 6 *